Federal Emergency Management Agency
United States Fire Administration
Emmitsburg, Maryland 21727

I0470846

The operation of an emergency vehicle is one of the most common functions performed by today's fire and emergency service organizations. Yet, it is also one of the most dangerous.

During the 1980's, 179 firefighter deaths were the direct result of emergency vehicle accidents, representing 15 percent of all firefighter fatalities. Of course this figure does not include the hundreds of firefighters injured in emergency vehicle accidents. The National Transportation Safety Board (NTSB) recently focused on the issue of emergency vehicle operations in its "Special Investigation Report on Emergency Fire Apparatus." The NTSB shares concern with the United States Fire Administration (USFA) and the United States Fire Service over this tragic and unacceptable loss of life.

Fart of the solution to the problem is an effective driver training program for fire and emergency service personnel operating emergency vehicles. This Emergency Vehicle Driver Training Package is designed as a resource manual for any fire department or emergency service organization wishing to develop or enhance an Emergency Vehicle Driver Training Program. This comprehensive package includes both an Instructor's Guide and a Student Workbook (which may be reproduced in order to provide each student with his/her own copy). It is a USFA reprint of a manual developed by the Volunteer Firemen's Insurance Services (VFIS), Inc., of York, Pennsylvania. The USFA is appreciative of the VFIS for allowing our reprinting of this manual.

The Federal Emergency Management Agency and the USFA are deeply committed to enhancing firefighter health and safety throughout the Nation. This Emergency Vehicle Driver Training Package has been published as part of our continuing efforts to provide useful information on health and safety issues to the fire and emergency services community.

It is our most sincere hope that all fire and emergency service personnel remain "alive on arrival."

THE AUTHORS
LOUIS J. KLEIN, ASSISTANT CHIEF

Lou's father was president of a rural volunteer fire company. Naturally, in his boyhood days Lou became quite familiar with the fire service and joined a company when he was of age.

Lou completed high school at York Catholic High School in York, Pennsylvania, and started his collegiate career at Shippensburg State College as a biology chemistry major. While transferring colleges in 1964, he was drafted. He then enlisted in the United States Army and continued his studies as a medical technician. He spent 18 months at Fort Detrick, Maryland, doing biological warfare research. He was discharged in December, 1968, and began employment at the laboratory of York Hospital. He resumed his education at York College of Pennsylvania, attaining his registry as a medical technologist. Remaining an active member and officer of a local volunteer fire department, he decided in the Winter of 1969 to take a Civil Service exam for a position of paid firefighter with the York City Fire Department. He was appointed to the position of lieutenant in September 1977, as executive officer of "A" Platoon and also the department training officer. During this year he completed the requirement for his degree in fire science from Harrisburg Area Community College. In July, 1980, Lou was appointed to the position of Assistant Chief within the York City Fire Department with the job responsibilities of platoon commander of "D" Platoon, Training Officer and Director of Photographic Services.

Lou has received numerous community service awards from such organizations as the Knights of Columbus, Optimist Club, Sertoma Club and Junior Chamber of Commerce. He has also received the Chiefs Commendation Award from the York City Fire Department as well as a Letter of Commendation from the Mayor of the City of York for his rescue of a man from a burning vehicle.

Lou also has worked as a part-time instructor with Harrisburg Area Community College, teaching in the fire science division. He has conducted numerous workshops at the Fire Department Instructors Conference in Memphis, Tennessee, and he is an instructor for Pennsylvania State University and has completed courses at the University of Maryland and Delaware State Fire Schools.

He holds membership in the following organizations: The International Society of Fire Service Instructors; The International Association of Fire Photographers; The International Association of Fire Chiefs - Eastern Division, of which he is a member of the Scholarship Committee; and Pennsylvania State Arson Investigators Association.

Lou came to Volunteer Firemen's Insurance Services in January, 1980, as part-time Assistant Director of Safety and Education under the direction of former Fire Chief Robert W. Little, Jr. In September, 1981, he joined the staff full-time. At VFIS, Lou's area of responsibility is to develop safety programs for the fire service and make those programs available to the fire departments insured by VFIS. He also lends assistance to the various agents and staff of VFIS, understanding the problems and needs of the fire service. Since the death of Chief Little in July of 1982, Lou has assumed the responsibilities of Director of Safety and Education at VFIS.

On June 1, 1984. Lou was promoted to Vice President of Client Support Services at VFIS. In this capacity, he is responsible for the programs, public relations and other services directed to the fire department and emergency medical organizations the Agency insures in 45 states.

THE AUTHORS
STEPHEN C. LANE

Stephen C. Lane received his Master of Science degree in Criminal Justice from Eastern Kentucky University in 1974. After attending the Durham Public Safety Officers Academy in 1975 in Durham, North Carolina, he developed an intense interest in Fire Protection. From 1975 through 1980 he was the Coordinator/Lead Instructor of the Fire Science Technology degree program at Harrisburg Area Community College, Pennsylvania, attaining the rank of Assistant Professor. During this period, he worked with various state and local agencies in a consulting capacity to promote Fire Protection. In 1980, after the TMI Incident, he served for two years as the Fire Brigade Training Coordinator for Three Mile Island. Duties included training plant personnel and off-site fire companies in response methods to protect the installation from fire. Since 1982, Lane has been the head of his own Fire Training Consulting firm. Since then he has served as Assistant for Emergency Services, Carlisle. Pennsylvania, and as Emergency Management Coordinator, Silver Spring Township, Cumberland County, formulating Emergency Response Plans and special projects. A Pennsylvania Certified State Fire Instructor, he is presently a private consultant to the fire service, industry, health care institutions and the nuclear industry.

FOREWORD

The value of driver manuals has long been recognized. During the last few years, however, the value of Emergency Vehicle Driver Safety programs has been evident through the tragic results of accidents involving Emergency Vehicles. The product of these occurrences has too often been serious injury, death and legal ramifications to citizen and emergency personnel alike. Attempts have been made to identify the problem, to isolate the common variable intrinsic to these accidents. Though the reasons span from roadway conditions to vehicle operation, the common denominator which has the greatest hopes of changing the condition is to provide driver/operators with a better understanding of their vehicle.

Emergency Service vehicles are a "different breed" due to being specially designed for specific uses and being operated during highly stress-filled operations. The fact of when they are used may have already altered the driver/operator's temperament. Within the vehicle's construction: shape, style, power and weight distributions are mechanisms creating unique limitations to use. Their initial cost and that of repair or replacement dictate a need for special driving abilities. For such reasons, driver/operators must be instilled with in-depth training as to proper operation through various ranges of performance.

This program has been offered in many locations across the United States and its intent is to serve as a means by which particulars concerning safe operation of emergency vehicles may be discussed and demonstrated.

The program deals with topics bearing on both the potential and past problems as cited in a number of studies. Information given shows where apparatus shortcomings may be found during operation and suggests means to operate within safe ranges. The program's philosophy is that a safe operator will not operate his vehicle beyond safe boundaryline limits to the point of "no return". The under current of this program is: safety through avoidance. There are incidents that should not have occurred, others whose results could have been reduced. Such cause and effect is also discussed. This program, however, does not teach skid recovery. It does teach skid avoidance. The methods here teach ways to function without entering the danger zone of operation.

Intended for driver/operators, the benefits are also realized by those who will serve as crews. Department Administrators are furnished with a means by which to gauge performance of present or future operator candidates, as well as to have available a training mechanism for new members. Such a program benefits drivers, their department, fellow members and the community by ensuring safe response.

PREFACE

Driver training. Of all the areas of emergency service training, this one tends to be the least popular and the one regarded as least important. This problem is usually related to a negative attitude. When the subject is brought up, we almost always hear one of the following excuses:

1. This particular type of training is not exciting or demanding as many of the other areas of emergency service training.
2. All of us can drive so why do we need driver training?
3. The cost of training involving emergency vehicles is too high; wear and tear on the vehicle is not worth the risk.

The purpose of this manual is to provide the training officer and those who are responsible within the department to train their vehicle operators with a better understanding of the seriousness of driver training. We hope to put into their hands an effective tool to accomplish these results. Hopefully, this text will be of interest to both the student and the instructor. For the student involved in an ever increasing number of emergency vehicle responses, we can shed a new light and a new seriousness and responsibility on his very important position of driving an emergency vehicle.

There is probably no situation in which you will be involved in the emergency services that will carry as much responsibility as the safety of your vehicle, manpower and the citizens around you.

For the instructor, we feel this will be a fast moving and dynamic driver training program that features our major concerns in the psychology of driver training.

The intent of this program is to stimulate the thought processes of the students and to make them aware of the tragedy, financial loss and the legal and moral responsibilities that they have when operating an emergency vehicle. We try to prove to students that they may not all possess the fine degree of coordination that they may think they have in operating an emergency vehicle. It is our philosophy that just as not every firefighter is capable of being a good interior firefighter and a good aerial person, not every EMT has the ability to progress to the stage of paramedic. We feel that every person, even though they have a drivers license, does not have the makeup to be a good emergency vehicle operator. It is our hope that this needless waste of money, manpower and machinery can be kept to a minimum. It was with this in mind that this program was developed.

FA-110 9/91

Instructor Manual

United States fire Administration

Emergency
Vehicle
Driver
Training

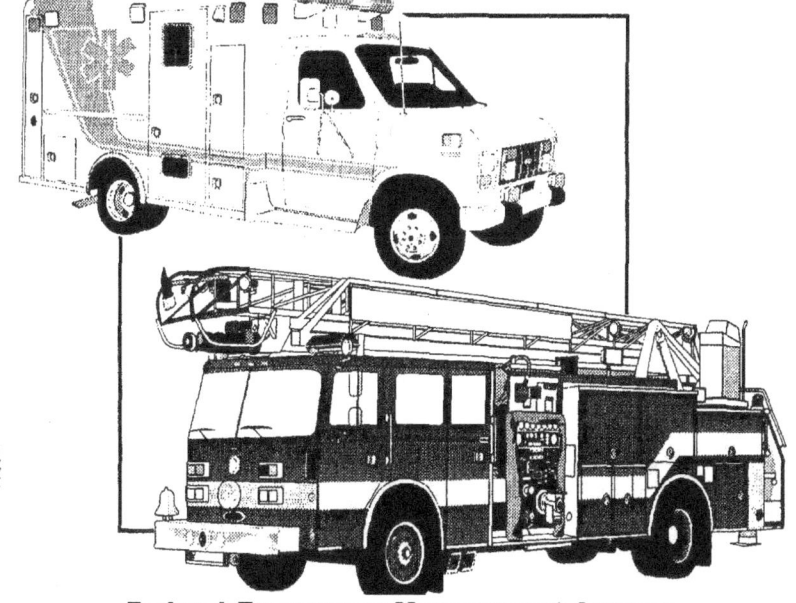

Federal Emergency Management Agency

COURSE OBJECTIVES

- To define the needs analysis for the necessity of an emergency vehicle operators course.

- To identify the problems facing the operators of emergency vehicles.

- To motivate emergency vehicle operators to recognize the importance of emergency vehicle operator training.

- To define the personal qualities and attributes of a candidate for operating an emergency vehicle.

- Review the legal responsibilities of the emergency vehicle operator.

- Discuss the physical forces involved in the operations of an emergency vehicle.

- To express the importance of a preventative vehicle maintenance and maintenance record programs.

- To review the necessity for standard operating procedures.

- To examine the state and local laws dealing with the operation of an emergency vehicle.

- To provide the student the opportunity to perform "hands-on" operation through a designated driving course.

TABLE OF CONTENTS

Page

I. Introduction   1

   A. Needs   1

   B. Risks   2

II. Identifying the Problem   5

   A. Family Dependence for Income   7

   B. Emergency Vehicle Accidents: News Articles   9

      1. "New Fire Truck Rams Tree; One Injured"   9

      2. "Fire Truck Crashes: Three Volunteers Hurt at Steelville Fire Scene"   11

      3. "Accident Claims Lives of Five Firemen*"   14

      4. "Fire Fighter, Volunteer Fire Company and County Sued in Vehicle
         Accident"   16

      5. "Fire Truck Overturns; 3 Injured"   18

      6. "Sayre Fireman Dies Under Wheels"   20

      7. "Fire Truck Hits Car; Couple Dies"   22

      8. "A Tragic Loss - A Tragic Record"   25

   C. VFIS Loss Statistics   26

   D. Indiana State University of Pennsylvania Study: Synopsis   28

   E. General Information   28

III. Motivation Exercises   32

   A. Background   33

   B. Schools of Thought Concerning Motivation   33

   C. Physical Need   34

   D. Mental Need   34

   E. Proof of Physical Need for the Course   36

   F. Proof of Mental Need for the Course '   37

IV. Personnel Selection   39

   A. Driving System Components   42

   B. Apparatus Driver Selection   43

      1. Human Aspects   43

         a. Attitude

         b. Knowledge

         c. Mental Fitness

         d. Judgment

         e. Physical Fitness

         f. Age

         g. Habits

         h. Driving Characteristics

  2. Acquired Ability   44
     a. Driver's Licence
     b. State and Local Laws
     c. Defensive Driving Techniques
   3. Vehicle Characteristics   45
     a. Type of Vehicle
     b. Components
     c. Special Training Required
  C. Personnel File   46
  D. Discussion Topics   49
  E. VFIS Personnel File Form   52
  F. Review of Factors Discussed   53
V. Legal Aspects   55
  A. Negligence   58
  B. Acting "in good faith"   58
  C. Legal Aspects of Emergency Vehicle Operations   61
   1. What is a True Emergency?   62
   2. What is Due Regard for the Safety of Others?   62
   3. Case History   63
VI. Physical Forces   67
  A. Types of Energy   68
   1. Potential Energy   68
   2. Kinetic Energy
   3. Velocity
  B. Types of Control   69
   1. Velocity Control
   2. Directional Control
  D. Physical Forces and EV Control   70
   1. Friction   70
     a. Tire and Road Friction   71
     b. Friction at the Brakes   72
     c. Velocity Control and Friction   74
     d. Possible Ranges of Pavement Drag Factors (Chart)   76
   2. Law of Inertia   77
   3. Law of Momentum   79
     a. Momentum and Inertia   80
     b. Law of Conservation of Momentum   81
   4. Law of Reaction   82
   5. Centrifugal Force   82
     a. Examples
     b. Principles of Centrifugal Force
   6. Centripetal Force   85
  E. Following Another Vehicle   85
   1. What is a Safe Following Distance?   86
   2. Estimating Following Distances   86
     a. What is Stopping Distance?   86
     b. How to Tell When the EV is Far Enough Behind?   87
     c. When Should Following Distance be Increased?   88
     d. Following Distance in the Emergency Mode   89

       3. Stopping Distances   90

          a. Light 2-Axle Trucks (Chart)   91

          b. Heavy 2-Axle Trucks (Chart)   91

          c. 3-Axle Trucks and Combinations (Chart)   92

    F. Seat Belts (Passenger Restraints)   92.1

       1. US DOT National Highway Safety Administration Observations   92.1

       2. OSHA Section 1915.100 Proposed Standard   92.2

       3. Crash Dynamics   92.3

          a. Car's Collision

          b. Human Collision

          c. Factors Contributing to Injury and Death

       4. Unrestrained Occupants   92.4

       5. Belted Occupants   92.5

       6. Myths vs. Facts Regarding Seat Belt Use   92.5

VII. Vehicle Maintenance and Records   93

   A. Enacting PM (Preventive Maintenance)   95

       1. Apparatus

       2. Equipment

       3. Personnel

VIII. Vehicle Standard Operating Procedures   103

   A. Reasons for Constructing

   B. Areas Addressed in SOP

 IX. Appendices

   Appendix A: State Laws/Local Statutes

   Appendix B: Scarsdale, New York, Ambulance S.O.P.

   Appendix C: VFIS Driving Course

   Appendix D: Instructor Key for Student Workbook

   Appendix E: Bibliography

 X. Overhead Transparency Templates

 XI. Index

SECTION I
INTRODUCTION

INTRODUCTION

The driver of an emergency service vehicle carries heavy responsibilities for the safety of his vehicle, his comrades, and other vehicles and pedestrians along his route. He must be constantly aware of these responsibilities and have his vehicle under control at all times. He must be familiar with the traffic laws, particularly those that apply to him and his specialized driving capacity.

He must recognize his limitations, and those of other drivers on the road and realize while he may know what he is doing, the other driver or pedestrian may not. Therefore, he must always be prepared for the unexpected to happen.

The emergency vehicle driver must possess fine coordination in controlling his vehicle and reacting to traffic problems. He cannot drive faster than traffic permits, nor should he drive faster than his ability to stop in an emergency. The right of way given to an emergency does not relieve him of his responsibility for the safety of all other users of the streets. The law allows certain exemptions to emergency drivers while responding with flashing lights and sirens, or similar devices. But it does not overlook any arbitrary use of these rights when such use endangers the life or property of others.

Excessive speed, reckless driving, failing to slow down or obey signals, disregarding traffic rules and regulations, and failing to heed warning signals are often prime factors in such emergency vehicle accidents.

Probably the key area in the whole problem of drivers and driver training is the first phase, which is the basic selection of EV drivers. Just as heavy a responsibility rests on the shoulders of those who select the driver. This selection generally fails on the commander of a firefighting or emergency service unit, the chief or other designated person. Before a person is selected to drive, he should be closely screened as to habits, aptitudes, limitations and, most important, his attitude. Only then, should he be permitted to enter a candidate course in driver training and only then should he be tested and trained for the job.

As we are well aware, the results of an emergency vehicle accident can vary in degree of severity from costing a few dollars and a few hours of downtime to a disaster, involving total loss of a vehicle, severe injury or death to emergency service personnel and/or civilians.

If we take a few minutes to analyze this impact more closely, we will find there are four areas with which we should be concerned:

1. The risk of personnel injury or death to ourselves or fellow emergency service personnel. This particular area can especially affect your organization if you are a small rural operation. Imagine, if you would, losing four or five of your key personnel by injury or death, whether their services are lost for a few days or forever.

2. The risk of injury or death to others. This includes pedestrians or other vehicle operators that your actions affect while operating an emergency vehicle. It would be a difficult thing to live with, if you, (knowingly but through carelessness or improper actions of emergency vehicle operation) were responsible for the death or injury of another fellow human being.

3. The loss of the emergency service equipment to the community. As we know, all our emergency vehicles are very specialized pieces of equipment, made in limited numbers. Therefore, repairs or replacement is a lengthy process. Most of us do not have, the luxury of having many

· **Personnel Injury or Death**
· **Peripheral Injury or Death**
· **Equipment Loss**
· **Long Term Impact**

similar pieces to utilize as backup vehicles. When a primary piece of equipment is lost to an accident it impacts the community and ourselves as department personnel. We have a financial expenditure to contend with whether or not we have insurance coverage to replace or repair the damaged vehicles. We also face the consequences of not being able to provide emergency services to our community, whether it be in EMS or fire protection, during the repair or replacement of this particular vehicle.

4. If we have caused injury or death to our personnel or civilian personnel, we have the problem of the long range impact to the families of this personnel. For example, if a victim of our accident is injured and taken away from his workplace, most insurance or compensation does not afford him the ability to provide for his family as he had while working in his occupation. In the case of death to an individual, we have the long term impact of not being able to provide for the care and welfare of the family, although there may normally be insurance monies to help these victims adjust to the loss of their loved one. We all know in this day and age the cost of living and raising a family. Consider this: how long would a $50,000 Death Benefit last your family?

SECTION II
IDENTIFYING THE PROBLEM

IDENTIFYING THE PROBLEM

To get a better grasp of the problems facing our emergency vehicle operators and what is happening, as far as accidents are concerned, we find it necessary to study different sources of information concerning accidents. The three cases that we will be concerned with are actual newspaper accounts involving emergency vehicle accidents. We will find, as we read through these reports of the incident, an element of human failure. Additional data to help us survey this important area will be found in actual loss statistics from the files of Volunteer Fireman's Insurance Services. And thirdly, government studies that have been completed concerning the problem with emergency vehicle operation will be presented. (A needs assessment for Pennsylvania in 1979-1980.)

(Instructor's Note-Read entire article of the following series of overhead transparencies and become familiar with the details as illustrated. Discuss with the student some of the highlights in each of the articles with emphasis on the following points: 1. Impact of the injury or death to the emergency service. 2. Impact of the damage to the fire department; vehicles and equipment lost. 3. Impact to the families of the injured or deceased emphasizing the following points:

Family dependence for income

A. Worker's Compensation

B. Insurance Benefits

C. Long Range Income Projections

4. Impact to the emergency service organization due to the reduction of manpower. 5. Impact to the community with loss of emergency services.)

Newspaper Article No. 1:

"New Fire Truck Rams Tree; One Injured"

Source: "The Oregonian" Portland, Oregon

Location: LaCenter, Washington

Date: Friday, April 17, 1981

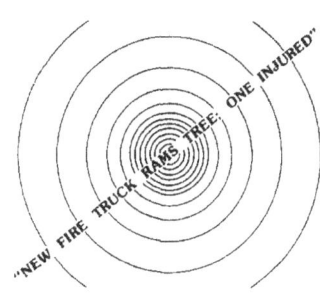

This engine, on its maiden run, was responding to a school bus fire. The bus was unoccupied. The engine was operated by a person not fully oriented to the differences between this piece and the company's old engine.

The travel route was over rain-slicked roads also involving a curve which was missed, leading to the accident.

Result of the accident was an injured driver, entrapped for more than an hour and extensive damage to the new engine.

Particulars:

1. The driver was the only person on the vehicle which the Chief stated was "unusual for only one firefighter to be aboard a truck when responding to a call." But, "She was the only one around."

    Had the driver waited for a crew to form? Having no crew available, was she attempting to make up for lost time during the wait?

    Did the fact of being alone deprive the driver of acting more carefully as may have been the case with an officer alongside to instill precautions?

2. The new engine was on its maiden run.

    Had the driver any knowledge of the similarities or differences between this piece and the old one?

    Was it thought that this new piece would handle as the old and negotiated the road in a similar manner?

3. The engine missed a curve.

    How familiar was the driver with the response route?

Had she driven other pieces on this route before under similar condi-

tions? Or was there still an element of unfamiliarity?

Newspaper Article No. 2:

"Fire Truck Crashes:

Three Volunteers Hurt at Steelville Fire Scene"

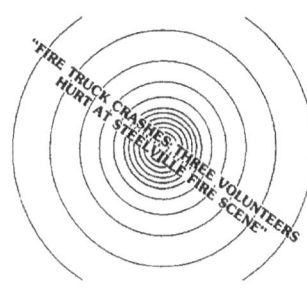

Source: "The Record"

Location: W. Fallowfield Township

Date:

1. The failure of brakes while traveling down a mile-long hill, after failing to negotiate a curve at 11:26 p.m., led to the injury of three volunteer fire-fighters.

2. The driver could not negotiate a curve, lost control of the piece and the brakes "went out" as the hill was descended. A collision occurred at the fire scene, at the base of the hill, with a stationary piece and the runaway continued west for 100 feet toward a house, where the engine went through shrubs. The driver chose the shrubs rather than an embankment.

3. The response was to a structure fire on Steelville and Bryson Roads, which resulted in damages in excess of $30,000 to the home. "Meanwhile, firefighters continued to fight the blaze at the 2 1/2 story house."

4. The accident resulted in a further response of seven ambulances to treat and transport the injured.

5. The 42 year old driver was the more seriously injured, while two other firefighters received minor injuries.

6. The Chief stated, ". . . We were lucky we had only one bad injury," and "You have to realize things like this do happen and we were fortunate things turned out the way they did. He (driver) did his best to avoid everything he could."

1. Visibility at this time of night, 11:26 p.m. may have interfered with operation. Terrain is rarely changing, with respect to hills and curves. Had the driver known of and anticipated the curve? What speeds were achieved?

2. Had the braking occurred before entry into the curve or while into it? Had brake fade occurred resulting in the total loss? Extreme momentum was achieved by a runaway situation evidenced by the distance traveled after collision.

3. Responses to dwellings are emergency conditions, its true. However, depending upon initial findings, fire scene updates and radio broadcasts such as; "Proceed at a reduced rate," for incidents which have been calmed, or "Proceed under caution," to warn of hazards to incoming apparatus, are furnished.

   Certainly, in all fairness, we do not know if the Fire Department piece was needed for a priority duty; search and rescue, sector control or water supply. Such is possible.

   Fireground Commanders and Command Posts should remember that the first priority is the "life hazard" at an incident. This concern should extend itself to responding units as well. Such is one of the many reasons for the establishment of a Command Post at the scene.

4. "Seven ambulances"

   Each additional piece responding emergency multiplies the chances of an accident, just as does each additional mile which must be traveled to a scene. This is a reason areas have performed time/distance studies and enacted provisions for the closest company to respond to an incident *along with* the company of jurisdiction. This has been used even if territorial boundaries must be crossed.

5. Did the driver's age or other characteristics bear on the incident? Was the driver qualified on this piece? If so, according to what requirements? Had

he consistently operated the engine without incident? What was different, if anything, about the piece's configuration to contribute to the accident?

6. Some may feel the Chiefs statements were fatalistic, as if Preventative Maintenance, training and other precautions could not have been taken to preclude the occurrence. If the driver had the presence of mind to avoid more serious ramifications, while in a runaway situation, could something have been done to completely eliminate the occurrence beforehand?

We give the Chief the benefit of the doubt, that the statements were out of relief that his people were not more seriously hurt, and others escaped injury, too.

Beware, however, of those officers who assume they are at the mercy of "the fates" or "destiny" and cannot/do not/will not anticipate such possibilities-and by so neglecting, place all drivers and accompanying personnel in danger.

"ACCIDENT CLAIMS LIVES OF FIVE FIREMEN"

Source: "The Pilot-News"

Location: Plymouth, Indiana

Date: Monday, August 2, 1982

On Saturday, July 31, 1982, shortly after noon, the lives of five of six men on the first of three engines responding to a house fire, were claimed. The response was to Lake Buzzard, six miles southwest of Plymouth. Declared "the worst accident in the history of Hoosier firefighting," it was the first time any Plymouth fireman had been killed in the line of duty since the department was formed in 1836.

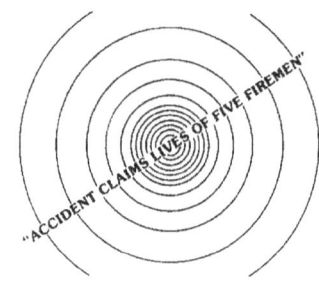

The accident occurred while the engine was heading south. It rounded a blind 90 degree curve, and as it did, it came across a northbound car ahead making a left-hand turn from the narrow single-lane highway.

It is believed the driver, upon seeing a car, swerved to avoid a collision, and the pumper's center of gravity shifted. The engine veered off the road, striking a guardrail, went down an embankment and flipped over. The engine came to rest on its back entrapping five of the six onboard. One of the three occupying the cab escaped serious injury when he was thrown through the windshield.

Initially, the cause of the accident was unknown. Skidmarks at the scene seemed to indicate that it wasn't the curve which caused the engine to go out of control. The County Police had the engine impounded for a thorough investigation to examine the 1965 piece for mechanical failure. A specialty laboratory performed the investigation and found no evidence of mechanical failure in; 1) suspension, 2) steering, 3) brakes, 4) engine drivetrain, or 5) tires.

Final ruling for cause was the action of the high center of gravity (hose bed, tank and water) during the evasive maneuver which caused the driver's loss of control and the subsequent overturn.

The result of this accident required the enactment of an "automatic mutual aid" system whereby closest of two county fire departments would respond.

Over twenty retired firefighters returned as volunteers and other departments offered manpower, equipment and other aid as might be required.

The loss of personnel in this accident bears stating. From NFPA acquired statistics provided to Firehouse Magazine by Chief Michael R. Hargreaves of the Mishawaka, Indiana EMS Department, it was learned that between 1978 and 1980, 83 firefighters lost their lives while enroute or returning from calls.

The following information is provided regarding the victims of the Volunteer Fire Department so the full impact of this tragedy is understood.

1. Victim I, 62 years old. A veteran driver of the department who had 17 years, 8 months of service. He was taken from the wreck at 12:30 p.m. by volunteer personnel in the second engine which had been following. He died in the emergency room of the hospital.
2. Victim II, 24 years old, a volunteer with 3 years and 41 days of service. Freed at 2:15 p.m., he died at 3:57 p.m. at Parkview Hospital.

The following were declared dead at the scene and removed after 3 p.m.;

3. Victim III, 32 years old, a volunteer for 6 years and 6 months.
4. Victim IV, age 26, a volunteer for the department, 22 days short of 3 years service.
5. Victim V, 29 years old, a volunteer with 200 days on the department.

The sixth member of the responding crew, who escaped with his life, was age 27. He was thrown through the windshield and escaped with minor injuries.

"FIRE FIGHTER, VOLUNTEER FIRE COMPANY & COUNTY SUED IN VEHICLE ACCIDENT"

Source: "Pennsylvania Fireman"

Location:

Date: May, 1983

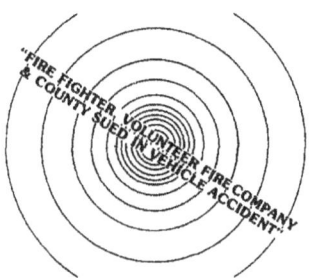

While responding to an emergency (red lights and siren) to a non-emergency washdown, an engine attempted to pass a car stopped in the eastbound lane, the engine's direction of travel.

In the westbound lane, two cars were approaching the engine. When the first westbound car stopped, the car behind skidded. At this point, the engine struck the skidding car driven by a 17-year old boy, crushing it against a bridge, killing the youth. The engine driver, an 8-year veteran, was removed from driving duty following the accident.

Parents of the victim have filed suit against the 1) County, 2) volunteer fire company, and 3) the volunteer driver, seeking damages in the sum of $24 million. The parents have requested a jury trial, claiming the following in their 8-count suit:

1. That the volunteer fire fighter driver was driving in a negligent manner, and charging him with:

    A. The failure to give full attention to the operation of his vehicle at a speed greater than reasonable and prudent.

    B. The failure to operate the vehicle within the proper lane of travel.

    C. The operation of his vehicle as an emergency vehicle when unnecessary to do so.

    D. The failure to maintain control of his vehicle.

E. The failure to avoid striking the vehicle being operated by the deceased.

2. The County failed to properly train and supervise fire fighters who operated emergency fire trucks and equipment prior to and on the date of the accident which "resulted in the deceased being deprived of his life and his right to travel safely upon the public roads."

Damages are also being sought by the victim's parents (plaintiffs) for the "great mental anguish, emotional pain and suffering" since experienced due to their son's death.

The plaintiffs' lawyer stated, "We feel that certainly they've been damaged that much. It's hard to put a dollar figure on the life of a 17 year old."

"FIRE TRUCK OVERTURNS; 3 INJURED"

Source: "Bucks County Courier Times"

Location: Bristol Township

Date: Friday, October 3, 1980

On the morning of Friday, October, 1980, an engine of No. 2 Fire Company was responding to a structure alarm at Penn Brush Paints on Route 13. This reponse was to assist the Porter Fire Company. Unknown at the time, the alarm was false, caused by a short in the alarm lines. Such false alarms had been received from this location before.

As the engine proceeded east toward Route 13, on Edgely Road, it collided with an auto at the intersection of Mill Parkway and Edgely Road.

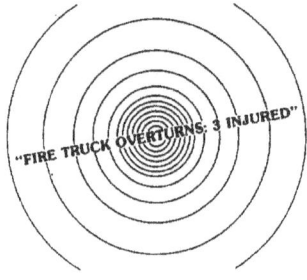

A witness had stopped her car to permit passage of the engine, but other cars passed her. Suddenly, a car behind her, shot past and entered the intersection. The engine swerved to the left to avoid the car, skidded across the road and was struck by the automobile. The overturn of the engine which resulted, was likened to a "dying elephant." This slow overturn was attributed with saving those aboard the fire engine from serious injury. Prior to initial impact, those firefighters riding the tailboard, saw the car enter the intersection and jumped off the engine. None of those "bailing out" were hurt seriously.

Other injured were: the driver, who was trapped in the cab along with the Officer. A firefighter suffered minor injuries in the jumpseat position behind the driver. Two months prior, he had received a hairline fracture to his back at a steel supply company fire.

As to the engine driver's actions upon entering the intersection, the firefighter stated, ". . . the driver slowed down, checking both ways while all

18

but stopping. The (engine) driver then "jumped on it" (accelerated), then saw the car. The car "hit the engine broadside and rolled us over."

The accident should not have occurred, one witness stated. "He (the driver of the car) had to see them (engine). He had to hear them."

Mr. Bob Jones, of the gas station adjacent to the accident scene estimated the car to be going 50 MPH and the engine to be going 45 MPH. However, witness #1 said the engine had slowed for the intersection and was not going fast when the accident occurred.

Another conflicting statement made by yet another witness was that the engine was swerving as it went through the intersection "at a nice clip."

The engine, a 1963 American LaFrance, valued at $93,000 was a total loss according to the Fire Chief. Due to another engine being out of service for brake work, the Fire Company must function with only two other engines.

Particulars:

From the report, it seems the engine approached the intersection and had the red light. The cars on Mill Parkway seem to have had the green light, passing witness #1 who has stopped. The witness's car was possibly seen as an obstruction to forward progress and passed.

Whether the engine "slowed," "all but stopped," or "was going a nice clip," is subject to speculation. However, the hazards of intersections to emergency vehicles may be seen with this accident.

Understanding the dynamics of physical forces, such as momentum during impact, as well as directional changes, the need to anticipate such occurrences stands out.

"SAYRE FIREMAN DIES UNDER WHEELS"

Source: "The Sayre Evening Times"

Location: Sayre

Date: August 14, 1976

On the evening of August 13, 1976, at approximately 10:35 p.m., a Lieutenant, aged 35, of the newly-formed Fire Company met death tragically. The Lieutenant's demise was not due to flashover, structural collapse or explosion, but to a mishap with an aerial truck and a mistake of judgment on his part.

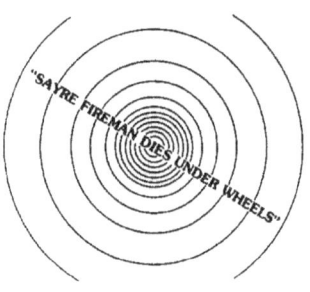

The Borough's largest piece, an aerial truck of Engine Co. 1, was responding to a car fire of minor scale along with other apparatus. It was found to be "nothing more than a minor electrical fire and there was no appreciable damage to the vehicle involved."

The Lieutenant attempted to join the response by driving his car to a location where the fire apparatus would slow to make a turn. He stood on the corner and waited until the aerial unit approached.

As the aerial slowed to make the turn from Keystone Avenue onto Center Street, the Lieutenant attempted to board the piece on the run. Losing his grip on a hand rail, he fell to the road, "rolling under the truck." He died at the scene as a result of the extensive internal injuries caused when the rear wheels ran over his body.

The Lieutenant's accident changed the lives of a son, two daughters, a brother and sister, mother, grandmother, several aunts and an uncle, as well as his nieces and nephews.

Particulars:

This accident drives home the requirement for caution and enforcing policies to prohibit such acts which may needlessly endanger personnel. Emergency service response to an incident denotes possible tragedy and loss at the scene. Acts which are performed, of a sacrificial nature, are admirable, if necessary. However, Emergency Service personnel are not gladiators, trained to be expendable. Their value to society and their partners is through a continued ability to protect the victimized.

No scathing remarks concerning this loss are warranted, for the loss of the Lieutenant tells its own story. It emphasizes the need to take precautions and to instill methods of conduct and behavior to ensure safety.

"FIRE TRUCK HITS CAR; COUPLE DIES"

Source: "The Palm Beach Post"

Location: Fort Pierce

Date: Wednesday, June 21, 1978

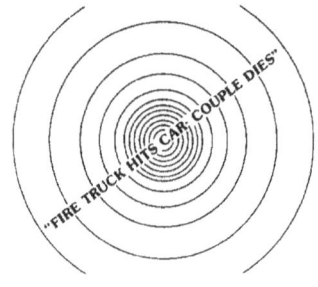

On Tuesday, June 20, 1978, Bill and Judy Jones were enroute shopping or to the new County Building, it is surmised. Mr. Jones (75) and his wife, Judy (73) had dated several years before their December marriage just six months prior. Mr. Jones, a retired barber, had difficulty hearing, being completely deaf in one ear.

At 9:40 a.m., this morning, a tanker responding to a house fire, drove through a red light at Orange Avenue and 25th Street, striking the Jones vehicle broadside. The tanker, carrying 2,000 gallons of water and estimated to be going 35 MPH, rolled on top of the car and then struck two other cars stopped at the intersection.

Mr. and Mrs. Jones died instantly. The firefighter/driver (32) was only slightly injured, as were the occupants of the two other vehicles.

State Highway Officers stated the Tanker Driver had slowed, increased his siren, but there was no way the driver of the Tanker could have seen the car coming. The Driver said, "I slowed down when I got to the intersection, but I didn't see the car."

Mr. Jones had driven "from behind another car and just seemed to have nothing else on his mind but making that green light." It was found that the windows on the Jones car had been rolled up, further compounding their ability to hear the siren.

According to Fire Department officials, "emergency vehicles have the right-of-way in an intersection, but 'must proceed with caution'."

Particulars:

1. The emergency vehicle had the red light, while the citizens had the green. Too often the reality of citizen drivers is conditioned by visible factors *normally* and *routinely* seen by them on the highway. Such includes; traffic lights, stop signs, speed limit signs, yield signs and painted arrows on the roadway.

   The intrusion of an emergency vehicle can momentarily stun or disrupt drivers unaccustomed to them. Such may also result in a total or temporary inability to properly react.

   Emergency vehicle drivers should remember that their reality may not be the same as a citizen's. Drivers of EVs should perform accordingly, exercising the proper caution.

2. Members of the Emergency Services are very often required to be physically qualified for membership in Fire and EMS organizations. Often, they are not, because no such requirement has been established, The rigors of the job require they be without restrictive handicaps.

   It should be remembered that the same stringent health requirements do not apply for a person to be a member of "society-at-large."

   Infirmities exist within the citizenry operating on the highways which may conflict with emergency response and cause erratic behavior. The reactions by such drivers may result in hazards to reponding personnel by involving emergency vehicles in accidents. Common among such traits which may threaten you are;

   A. Drug and alcohol use/abuse

   B. Psychological sets:

      1. Egoism:, being self-centered to the point of not giving way.

      2. Suicidal

      3. Altruism: unselfish concern of others. This type of person may tailgate and follow to assist at the scene.

4. Immature

C. Physical Infirmities:

1. Heart conditions

2. Hypertension

3. Handicapped: requiring specialty arrangements or methods built into the vehicle to drive it.

4. Hearing problems

5. Sight problems

One or more of the above, or other conditions, may exist. If they do, the normal driving "realities" have been altered thereby lending potential problems to the Emergency Vehicle on the highway.

You may not recognize which, if any, of the above factors exist on the highway with you. You may, however, anticipate their existence and respond accordingly to provide for your safety and that of others.

Newspaper Article No. 8:

"A TRAGIC LOSS-A TRAGIC RECORD"

Source: "The Maryland Fire & Rescue Institute Bulletin"

Location: University of Maryland

Date:

Late last month a 44 year-old career firefighter and a father of three died in the University of Maryland's Shock Trauma Center of injuries received in an apparatus accident three weeks earlier. His death was a tragic loss, especially so when one takes a closer look at the incident.

The early morning accident occurred when an engine company and a truck company collided with each other while responding to a street box. Even though both pieces of apparatus were stationed in the same house, they were approaching the scene from different directions. Six other firefighters were injured in the collision which totally destroyed the 1973 pumper and heavily damaged the late model tractor-drawn aerial. Replacement and repair costs are estimated to be over $140,000. The box the units were responding to turned out to be a false alarm.

Since the beginning of 1979 there have been an alarming number of serious apparatus accidents in Maryland. Some examples:

1. A brand new tanker making its first response, flipped over when it was involved in an intersection accident. A civilian was seriously injured, the vehicle's 2,000-gallon stainless steel tank destroyed, and the chassis seriously damaged.

2. A heavy duty rescue squad (responding to what turned out to be a faulty automatic alarm) was hit at an intersection throwing the driver out of the cab. He escaped serious injury by hanging on to the cab's open door as the driverless squad continued down the street, jumped a curb and plowed

into a jam packed apartment complex parking lot damaging numerous cars. The accident sent the entire crew to the hospital and caused extensive damage to the squad, knocking it out of service for an estimated six to eight months.

3. A tractor-drawn aerial, belonging to the same company, was involved in another intersection accident a month later causing major damage to the unit.

4. In another accident, an ambulance following a medic unit collided when the medic unit stopped short after missing a turn.

All of these accidents, plus a number of less serious ones, have occurred in less than 90 days!

An important training axiom states that all emergency operations should be performed with three objectives in mind: effectiveness, efficiency and safety. An apparatus involved in an accident while responding defeats all three of these objectives . . . and endangers our own lives and property as well as those of the public we are trying to protect.

VFIS LOSS STATISTICS

The main issues learned from these accident studies are the deficiencies in the handling of the larger vehicles (i.e. pumpers and tankers) and the inability of ambulance drivers to maneuver their vehicles safely through intersections.

Intersection calamities are generally the most costly due to the nature of the accidents--broadside collisions. Here, more than all other accidents, exists the potential for loss of life and total loss of equipment.

Ambulance broadsiding is prevalent as the driver assumes the right of way, even with a stop sign or red light, under extreme emergency conditions. The ambulance is thus vulnerable to contact from cross traffic that fails to observe or heed the warning lights or sirens. This is true for the other vehicles as well. In addition, there exists a great deal of uncertainty of action at intersections where time and patience are needed to handle the situations safely. The driver seldom has these, or seldom perceives that he has these. The first priority is a speedy arrival at the emergency site, and he endeavors wholeheartily to do so, often with legal exemptions, yet precluding safety. The trade-off between speed and safety is of vital importance.

A second main issue concerns inadequate handling of pumpers and tankers, where poor maneuvering accounts for approximately 45% of all the accidents for the two. The accidents are attributable to the size of often unwieldy nature of the vehicles. Due to configurations required for emergency vehicles being necessary or unavoidable, skilled drivers are then required to operate them. Experience and training should decrease incidents such as "truck striking parked cars," "rigs backing into walls and vehicles," and the "emergency service personnel hitting rails and trees enroute" to the fire or emergency. Accidents such as these account for 28% of all the accidents and a high percentage of the cost. Also, the private vehicles of emergency service personnel share the relative difficulty at intersections, as do ambulances.

If the problems in handling vehicles and crossing intersections can be alleviated, a significant amount of money and number of lives can be saved. Emergency service personnel must learn to be as conscious of safety as they are of the need to reach the scene of the emergency quickly.

(Note to instructor)-A series of overhead transparencies illustrating type of vehicle and area frequency from actual loss statistics to be explained.

V.F.I.S. STATISTICAL INFORMATION

	PUMPERS/ TANKERS	AMBULANCE RESCUE SQUADS
INTERSECTION	16%	23%
INADEQUATE CLEARANCE – Fixed Object	45%	22%
INADEQUATE CLEARANCE – Moving Object	10%	18%

A STUDY CONDUCTED BY INDIANA UNIVERSITY
OF PENNSYLVANIA

This study was the result of the Pennsylvania Emergency Health Services 1980 Highway Safety Plan. The goal of the study was to prepare individual drivers of emergency vehicles to operate their vehicles safely in emergency and non-emergency situations. With a thorough review of all accident data, the following facts were investigated and determined for the Commonwealth's Emergency Health Services:

1. Does a high rate of accidents exist?
2. How extensive is the problem?
3. What is the most feasible and/or economical method of solving and counteracting such a problem, should one exist?

GENERAL INFORMATION

At the time of the study there were 1,079 known agencies throughout the Commonwealth of Pennsylvania that provide emergency medical services for the State. These known service units accounted for 1700 ambulances used to provide needed emergency medical services. There were, at that time, 24,000 registered EMTs and an estimated one or two ambulance attendants for each EMT. Using these figures, there were anywhere between 24,000 and 72,000 potential ambulance drivers in the Commonwealth of Pennsylvania.

Emergency Medical Technicians are required to successfully complete a basic standardized training course which is a minimum of 94 hours in length. Of these 94 hours, only three hours are devoted to operations, and of these 3 hours, approximately 1 hour is devoted to the driving task. On the other

**TOTAL NUMBER OF REPORTED ACCIDENTS
FOR FIRST AND SECOND YEARS**

◇ FIRST YEAR	◇ SECOND YEAR
202 Ambulance Accidents	226 Ambulance Accidents
27.9 Accidents/100 Motor Vehicles	29.7 Accidents/100 Motor Vehicles
28.4 Accidents/Million Vehicle Miles	31.1 Accidents/Million Vehicle Miles

**VEHICLE INVOLVEMENT FOR
FIRST AND SECOND YEARS**

	FIRST YEAR	SECOND YEAR
	201 Accidents ◇	210 Accidents ◇
Single Vehicle	106 (52.7%)	112 (53.3%)
Multiple Vehicle	92 (45.8%)	90 (42.9%)
Pedestrian	3 (1.5%)	8 (3.6%)

**AMBULANCE USE DURING TIME OF ACCIDENT
FOR FIRST AND SECOND YEARS**

Type of Use	FIRST YEAR	SECOND YEAR
	193 Accidents ◇	219 Accidents ◇
Responding to a Call	55 (28.5%)	60 (27.4%)
Transferring Patient	59 (30.6%)	65 (29.7%)
Returning from a Call	48 (24.9%)	56 (25.6%)
Other Use	31 (16.1%)	38 (17.4%)

**LIGHT CONDITIONS AT TIME OF ACCIDENTS
FOR FIRST AND SECOND YEARS**

Light Conditions	FIRST YEAR	SECOND YEAR
	189 Accidents ◇	212 Accidents ◇
Dawn/Dusk	19 (10.1%)	23 (10.8%)
Daylight	115 (60.8%)	108 (50.9%)
Dark	34 (18.0%)	58 (27.4%)
Unknown	21 (11.1%)	23 (10.8%)

**ROAD SURFACE CONDITIONS AT TIME OF
ACCIDENTS FOR FIRST AND SECOND YEARS**

Surface Conditions	FIRST YEAR	SECOND YEAR
	196 Accidents ◇	213 Accidents ◇
Dry	97 (49.5%)	130 (61.0%)
Wet	26 (13.3%)	22 (10.3%)
Snow/Ice	27 (13.8%)	28 (13.1%)
Muddy	0 (0.0%)	1 (0.5%)
Unknown	46 (23.5%)	32 (15.0%)

**TYPE OF COLLISION FOR
FIRST AND SECOND YEARS**

Collision With	FIRST YEAR	SECOND YEAR
	190 Accidents ◇	222 Accidents ◇
Another Vehicle in Transport	62 (32.6%)	61 (27.5%)
Parked Vehicle	32 (16.8%)	46 (20.7%)
Fixed Object	71 (37.4%)	76 (34.2%)
Pedestrian	3 (1.6%)	8 (3.6%)
Animal	4 (2.1%)	5 (2.2%)
Non-Collision	0 (0.0%)	8 (3.6%)
Other	18 (9.5%)	18 (8.1%)

**ACTION OF THE AMBULANCE AND ITS OPERATOR
AT THE TIME OF THE ACCIDENT
FOR FIRST AND SECOND YEARS**

Operator's Actions	FIRST YEAR	SECOND YEAR
	157 Accidents ◇	174 Accidents ◇
Backing Vehicle	60 (38.2%)	61 (35.1%)
Turning Vehicle Around	16 (10.2%)	21 (12.1%)
Approaching Intersection	10 (6.4%)	9 (5.2%)
Proceeding Straight/ Intersection	28 (17.8%)	30 (17.2%)
Turning/Intersection	14 (9.0%)	11 (6.3%)
Entering/Exiting an Expressway	1 (0.6%)	1 (0.6%)
Passing	13 (8.3%)	24 (13.8%)
In a Curve	10 (6.4%)	11 (6.3%)
Parking	3 (1.9%)	3 (1.7%)
Other	2 (1.3%)	3 (1.7%)

**CONTRIBUTING FACTOR ASSOCIATED WITH THE
AMBULANCE AND/OR OPERATOR FOR ACCIDENTS
DURING FIRST AND SECOND YEARS**

Contributing Factors	FIRST YEAR	SECOND YEAR
	157 Accidents ◇	183 Accidents ◇
No Violation or Failure	107 (68.2%)	122 (66.7%)
Skidding/Loss of Control	18 (11.5%)	16 (8.7%)
Failing to Yield	17 (10.8%)	19 (10.4%)
Improper Passing	7 (4.5%)	8 (4.4%)
Following Too Closely	5 (3.2%)	6 (3.3%)
Too Fast for Conditions	1 (0.6%)	5 (2.7%)
Reckless Driving	1 (0.6%)	4 (2.2%)
Misjudging Driving	1 (0.6%)	1 (0.5%)
Vehicle Failure	0 (0.0%)	2 (1.1%)

hand, ambulance attendants receive no standardized training program, which could mean very little or no training at all in relation to safe performance of the driving task.

(Notes to Instructor) At this time a series of overhead transparencies of the statistical information collected will be discussed.

In conclusion, highway safety program standard No. 11, Emergency Medical Services, emphasizes the necessity for safe operation of emergency vehicles and supports the requirement for providing drivers with the appropriate instruction. At the present time, in Pennsylvania, there is no standardized training program for ambulance drivers. At minimum, a Class 1 license is required and there is no additional endorsement or certification needed by the operator. For many drivers, the only formal education that they might have had in relation to the safe performance of driving is a high school driver education program or perhaps a driver improvement program, such as the National Safety Council's Defensive Driving Course. Some of the emergency services that responded to the survey did indicate that they provided both education and training in relation to the safe and efficient operation of the ambulance, but these programs were very minimal and there was no uniformity among them.

In summary, by studying the newspaper surveys, the VFIS Loss Statistics, and the Needs Assessment Study conducted in Pennsylvania, we find that much more can be done in the area of driver training. We have summarized the following points also concerning these studies:

1. Not all of our emergency vehicle operators have the necessary skill and ability to operate emergency vehicles. This leaves a tremendous amount of responsibility to the commander of the ambulance or fire corps. The commander must be more conscientious in surveying driver candidates

to approve or reject them due to ability, and other factors. (These factors will be discussed in the 'Human Aspects of Vehicle Operation'.)

2. Most of our vehicle operators are in such a habit of operating smaller vehicles on a daily basis, they may not consciously be aware of the difference in size, mass, and other features of the emergency vehicle that they are about to operate. Many may operate emergency vehicles in the same manner they operate their personal vehicles. This fact emphasizes the need to mention the areas of vehicle dynamics in a driver training program including the laws of physics and motion. Differences in operational characteristics and required traits may be discerned when these areas are studied.

SECTION III
MOTIVATION EXERCISES

This section deals with the physical and mental needs for such a program. Before going farther, something requires saying about motivation. It is this aspect which bears heaviest on the behavioral traits within each of us.

The definitions concerning motivation will give insight to those things which shape our actions and behavior.

MOTIVATE: to furnish with a motive or motives;

              to give impetus to;

              to incite;

              to impel.

MOTIVE: Some inner drive, impulse, intention that causes a person to do something or act in a certain way; an incentive; a goal.

MOTIVATION: Motivating or being motivated.

As may be found in the derivation of Motivation, it appears that we are affected by both internal forces, external forces and/or a combination.

There are two (2) schools of thought concerning Motivation:

1. One school of thought holds that a person must inwardly possess the desire to do something. That, no matter how convincing external reasons are, if the person does not desire to act or perform in a certain manner, no amount of external force will affect the performance. (Individual creates the results.)

2. The second school reasons that the human being is the result of many forces acting upon him and how the person views himself within his sphere of existence. It gives way to recognizing that the person acts according to physical and mental realities. (Forces create the results.)

We will explore two areas of need for this program: Physical Need and Mental Need.

PHYSICAL NEED

This aspect deals with a recognition by the individual of the probability and susceptibility of being involved directly or indirectly in a vehicular accident.

MENTAL NEED

This addresses the realization that we are or may become subject to behavioral traits or conditions which causes us to act in specific ways. The understanding of these needs and the forces creating them may do much to positively adjust or change our actions. This is a form of behavioral modification. It is something which we all need to varying degrees throughout our lives to smooth-out the "ruts" in which we may find ourselves. It is also a very good way to preclude the possibility of tragedy occurring to ourselves and others.

That is the "motive" of this section-to show you what changes may be required and to move toward such change.

Your "motive", this inner drive causing you to act in a certain way, may be due to traits acquired over the years due to routine, comfort and confidence.

1. Routine: "We've always done it this way," or "There was never any SOP or statement as to how it was to be done. My way was never questioned, so it's continued this way."

2. Comfort: "This is the easiest manner to accomplish what has to be done in the least amount of time. We are an emergency service, after all."

3. Confidence: "I know the way I'm doing this may not be the safest, but I know I can get it accomplished this way. When all else fails, I will use the best-known methods which have never failed me before."

What this program will show you is that the same characteristics of; (1) routine; (2) comfort; and (3) confidence may still exist, but there is a safer way to do things. There is nothing wrong with developing and implementing new methods and still retain a sense of routine (in a safer manner), comfort (knowing you are doing something as safely as possible), and confidence in the new methods because they work.

What is the inner drive or goal 'each of us has as drivers of Emergency Vehicles? How about some of these:

1. We are each responsible for the safety and lives of the citizens we serve. Can we be less dedicated to the safety of our fellow service partners?

2. Dynamic forces are at work in the world around us, impacting on us as soon as a wheel rolls. There is also a multitude of unexpected situations which may arise to conflict with our response. We have viewed some as weather, terrain, road conditions and mechanics of the vehicle. What we are attempting to do, through instilling proper action, is to circumvent the chances of the expected (over which we have some control) and the unexpected, to act against us.

3. The legal aspects of an accident may impact not only on the individual but also on the parent organization and governmental body. The ramifications can socially or economically ruin a member of an Emergency Service and bring discredit upon the Department and the Service as a whole. It may even cause the governmental entity to react by creating restrictive rules and regulations, a possibility within its power.

Let us take a closer look at the Physical and Mental Needs for this program; and you determine if motivation toward change may be required.

PROOF OF PHYSICAL NEED FOR THE COURSE: "N.A.P.D.

Would all of you please stand up? It is time for everyone to stand and stretch. (Allow students to stand for about one minute. Now begin the exercise by asking the following:)

1. Would those of you who have had an automobile or other type of vehicular accident in the last year please sit down? How about the last two years, please sit down?
2. Would those of you who have had an accident in the last five years please sit down?
3. Would those of you who have ever had an accident, or have been involved either as a driver or a passenger, please sit down?
4. (If there is anyone left standing, say:) "Would anyone who has had a member of your immediate family ever involved in an accident, please sit down?"

(You may then say,) "Do we have to say anything more about the proof of physical need for this course?"

* National Academy of Professional Drivers, Dallas, Texas.

36

PROOF OF MENTAL NEED FOR THE COURSE:

For the next exercise, proof of mental need, I am going to ask you a series of questions and I would like you to give me a rapid, one-word answer. I need everybody to answer out loud.

For example:

If we were going fishing on a lake, what would the device be called that we would fish from? The answer we would expect is "boat".

Now, would the class, as a whole, please give me a very rapid answer for the questions I am going to ask?

What do we call a funny story: (Answer: "Joke")

What do we call the white stuff that goes up a chimney? (Answer: "Smoke")

What do we call the thing around the neck of an ox? (Answer: "Yoke")

What do we call the white of an egg? (Answer to Expect: "Yolk")

Of course, the yolk of an egg is yellow and not white. The white is albumen. What we have done in a very short period of time is put you into what is called a "Psychological Set", or just a plain mental rut.

If we can do this, just think of the mental rut that our fire apparatus has been putting us in since the day we first started to drive on the job. Do we need to say anything more about the mental need for the course?

SECTION IV
PERSONNEL SELECTION

On September 13, 1899, Henry Bliss stepped from a curb in New York City and was hit by an electric horseless carriage. On September 14, he died and became the first traffic fatality in the United States.

Since that time we have become much better-actually we have become quite professional. We now kill over 50,000 people a year with the automobile.

We don't need driver training! All of us here know how to drive. That is our American birthright. Then why do we kill and maim so many people per year with the automobile?

How many of you were asked if you could climb a ladder when you joined the department? Whether you did or not, you received many hours of training to make you accomplished at the task. How many of you received any driver's training? How many times have you been required to climb two or three stories up a ladder? How many calls require the use of some sort of vehicle? Nearly every call you will make in your career will require the use of some sort of motorized vehicle, and yet the ratio of training for this metal monster is low when compared to other job functions in the firefighting profession.

Since our aim is to achieve safe vehicle operation, thereby reducing or eliminating unwarranted accidents, we must view the main components of driving safety. Each component may be viewed separately and then jointly to see the dynamics and interrelatedness of the total driving system.

The individual parts of this driving system are:

1. The driver of the apparatus and those aspects which may influence performance.
2. Regulatory requirements which have been established governing acceptable behavior.
3. The characteristics of the vehicle which is the controlled aspect of driving safety.

These aspects have been dealt with before by other Fire Service Educators in the context of the "Man/Machine Concept." As you will note, the driver and his vehicle typify this concept in the most dramatic of senses.

The result of recognizing this interrelationship culminates, for the Fire Department Administrator, in the ability to record data pertinent to the safety of his organization and its personnel. From the driver's aspect, records of vehicle worthiness are retained, providing a means to determine if the equipment is safe for all who may use it. Such would be found in the Equipment or Maintenance Records.

From the standpoint of worthiness of drivers, the Chief Administrator has a record of drivers based on learned and demonstrated competence. Should drivers require additional or updated driver training, the records will indicate this. Also, training required in specific areas may be viewed. All in all, the creation and continuation of such records enables the Administrator to keep abreast of needs within his organization. Selection of drivers will be by established requirements for operating apparatus.

APPARATUS DRIVER SELECTION

HUMAN ASPECTS

· Attitude
  · Knowledge
    · Mental Fitness
      · Judgment
        · Physical Fitness
          · Age
            · Habits
              · Driving Characteristics

HUMAN ASPECTS: These ingredients are those provided by the prospective driver. They are the product of the constituent parts of this individual and how they will influence his abilities in either a positive or negative sense.

These Human Aspects have been categorized as follows for discussion:

1. ATTITUDE: The driver's disposition toward driving.

2. KNOWLEDGE: This is defined as a clear perception of a truth, fact or subject.

3. MENTAL FITNESS: The state of mind while driving.

4. JUDGMENT: The ability to make the right decisions.

5. PHYSICAL FITNESS: The lack of physical impairments which may jeopardize the driver's ability to maintain full control of the vehicle in the proper manner according to its design.

6. AGE: (This has been coupled with Physical Fitness.) Is the age of the driver such that the potential exists to have any of the Human Aspects impaired.

7. HABITS: A characteristic produced by constant repetition of an action.

8. DRIVING CHARACTERISTICS: These are the manipulative skill abilities.

A person may be affected in either a positive or negative manner in one or several of the above areas. And even the most positive person may be affected negatively by factors outside of his control.

Take, for example, the section on mental fitness. Studies dealing with Biorhythm have found we each have our "peaks" and "valleys" in our daily lives. Somedays we are "up," ready to react favorably to various forms of

See App. A-3 thru A-8

adversity or conflict. Other days we may be "down," often feeling at the mercy of conditions.

The information dealing with physical fitness and age shows that age, in itself, may be indicative of certain specific changes or traits. However, these characteristics, once recognized, may be countered, if the proper methods are applied. Impairments may exist, yet methods do exist to reinforce the positive capabilities within each of us and remove negative influences.

It is toward such recognition and the ultimate elimination of such negative aspects that this section is directed.

ACQUIRED ABILITY: These are abilities learned and demonstrated to agencies which sanction your performance. These agencies have established a "minimum acceptable criterion" of performance. They have drawn the boundaries between unacceptable and acceptable. They are regulatory in nature and philosophy.

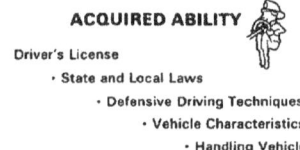

ACQUIRED ABILITY
Driver's License
· State and Local Laws
· Defensive Driving Techniques
· Vehicle Characteristics
· Handling Vehicle

1. Driver's License: Here a knowledge and ability must be proven or the granting agency may withold the privilege for you to operate a vehicle. And, a driver's license is a privilege-not a right as often mistakenly thought.

2. State and Local Laws: Establishing the methods by which a vehicle must be operated within a given locale or such privileges, once granted, may be withdrawn.

3. Defensive Driving Techniques: These acquired abilities must be long term. Depending upon circumstances, they must be changeable and adaptable. They include:

   A. Reaction Time

   B. Braking Distance

   C. Stopping Distance

   D. Physical Elements (constantly at work upon driver and vehicle)

VEHICLE CHARACTERISTICS: This is the "Machine" portion of this dynamic partnership. The following categories are viewed as they relate to safety:

1. Type of Vehicle: This will indicate its potential hazards if not properly controlled. Each vehicle has its own idiosyncrasies and required method of operation. Operating outside of engineered ranges can spell disaster.

2. Components:

    A. Engine: What type of power does the vehicle possess and how is this power different from what has been used by the organization in the past?

    B. Drive Train: Do certain driving conditions indicate a change to operate the vehicle in various terrain?

    C. Braking System: Are special deceleration requirements applicable to maintain constant control of the vehicle. What may happen to vehicle and crew if less than what is required is applied.

3. Special Training Required of Drivers: This translates into competence-based training for driving under given conditions which may be experienced by operators.

    A. On-Highway Training: Are the abilities of the driver more than those required by other apparatus within the organization.

Do such differences exist that driving this piece demands special abilities and attention to defined details of operation?

B. Off-Highway Training: Is a realignment of the vehicle's components required? Must such be accomplished in order to retain control? Are certain modes of operation and types of terrain expressly forbidden by the manufacturer because they are outside the realm of the vehicle's capabilities?

PERSONNEL FILE

The results of Human Aspects and Acquired Abilities should be available in a centralized place in the organization. Such a place is the standard Personnel File (See VFIS Record Keeping System). This file contains the qualifications of each person within the department. It also serves as a means to select drivers, indicating required training, and for qualified drivers, training updates needed.

While containing pertinent data on all personnel as far as background, phone numbers and specialties available to the department, the file shows training obtained. These training programs may have been merely "attended" or those which were "successfully completed." There is a difference between the two types, the latter indicative of a competency-based learning process.

Many areas offer training solely as a means of information transmittal. There are no tests required nor demonstrations by participants of skills learned. This type of training may or may not have instilled abilities in participants. With no way to measure learning, the acquisition of knowledge and capabilities becomes an unknown.

Programs, such as this one by VFIS, will result in learned information, and also improved driving techniques by applying, in the field, aspects learned in the classroom. The proof of acquired abilities will be shown through competency-based testing.

46

Company/Departmental Personnel Files should never be taken lightly. The Administrator or Officer should not only know all of his personnel's abilities and limitations, but also situations which may exist which could produce detrimental driving habits. Review the documentation and, as Administrator, don't be hesitant about talking with your people to discover if something is happening to these individuals which you might be able to correct yourself or refer to others to resolve.

The Personnel File should include complete information on each person. Once put into use, it may stand as a legal document which may be used in court. Just as dangerous to the Fire Department Administrator is the total absence of such documentation. Since you understand the meticulous scrutiny to which such documents may be subjected, do not dismiss their use by saying; "Well, then we just won't keep Personnel Files." This can only bring you hardship and embarrassment as well as make your organization appear to be slack, unprofessional and less than what it truly might be. Also, questions which may require documentary proof, which such records would provide, may be unsubstantiated without the records to support the answers.

The solution? Establish an Administrative process for the department. If one exists, review it and purify it. Assure that information is maintained on personnel and equipment which indicates the truths of each area. If unfavorable aspects are discovered, correct them in a timely fashion. That's another reason for the records-to serve as a means of checking and correcting shortcomings. Use the forms available from VFIS as your adopted format with which you will comply. Such adoption provides an accepted means of documentation since these forms are designed to include information required to run a smooth, professional organization.

Would you believe there are Fire Department Administrators who do not know; what training their members have obtained, what physical or mental problems exist in the organization, if proper maintenance has been performed on equipment, what equipment has been issued, and very little about their personnel?

Don't those, whose safety will be entrusted to others, have the right to expect they will get safely to and from an emergency? As Officer, wouldn't you want to know as much as possible about those to whom you are entrusting others' safety?

Here are some questions you would want answered concerning the department's driver:

1. Is he physically capable of doing the job? Are there limitations to being an Emergency Vehicle driver?

2. Attitudinally, what philosophy does the driver subscribe to with respect to how he sees his duty? Is he mature enough to handle the job? Or, "I get there any way I can. Let'em hang-on." Or, "I am morally responsible for each person on-board and must drive defensively with this in mind." (With which driver would you feel safest?)

3. Past driving record. Behaviors can change-a negative record may have been due to inexperience, brashness of youth or circumstances. But how is the person behaving now? What exists on his personal driving record which could be duplicated within the organization with more horrendous results?

4. Is the person being treated for drug or alcohol abuse? How about for other conditions which may impair judgment or function? And, if not being treated, do symptoms exist which should be examined more closely toward driver fitness?

The majority of these questions may be answered by the company or department investigating committee when the person joins the organization. Updates are required throughout the person's membership. Nothing is static;

48

conditions and people change. It is the person's reaction and adaption to such change which must be monitored to ensure safe driving.

The Personnel File, therefore, provides a means of updating changes, and to identify and remedy potential problems before they occur. We do as much for hardware and apparatus to guarantee its longevity and service to the organization. Much more is required for our personnel.

DISCUSSION TOPICS

As a review, the following subjects are furnished. You decide how each of the topics may have a negative effect on an Emergency Driver's capabilities to perform safely.

TOPIC	SOME ANSWERS
1. Attitude	Immature. He's the only one whose safety he cares about or provides for. Brazen or show-off, more concerned with image than reality. Laid-back, but so much so that reaction is days late. Comic. Doesn't panic but sees humor in everything, from pedestrians scattering to personnel riding the tailboard like a water skier.
2. Knowledge	Some drivers don't even know their pumps . . . couldn't pour water out of a boot if the instructions were written on the heel. Does he know his piece well then?

3. Mental Fitness

Brick short of a full hod. May be okay for hydrant duty but to control propelled tonnage down a busy road going to an emergency? How has his mind been acting lately? Is he talking and acting differently, taking chances, acting in an erratic fashion with regard to his driving?

4. Judgment

Is he offensive rather than defensive? Check his attitude along with this.

5. Physical Fitness

Has he had recent medical attention? Is he back "to serve" too soon after an operation? Has his SCBA on the seat beside him been replaced by an oxygen cylinder so he can take a snort when he feels faint? Does his stomach have to be greased first or the "Jaws" used to get him in behind the steering wheel?

6. Age

Does this one get the crew rolling off the apron by yelling "Charge!!!"? Or does he curse a slow engine start by damning the Magneto (how many years since they've used them?)

7. Habits

Creates a horror show with siren and lights, closes his eyes as he floors the accelerator to get outside the station? Blasts into traffic as if all other vehicles already there will evaporate like fog and disappear?

TOPIC	SOME ANSWERS
8. Driving Characteristics	Take corners like a Daytona 500 champion, the devil with load shifts and spilled equipment. (Isn't that why other -engines are responding-to supply the equipment dropped?) Arrives at the scene with all the flair of the last Kamikaze assault against the U.S. Pacific Fleet. Likes the feel of power which is shown in his bulging eyes, grit teeth and swept-back hair created by a G-Force of 3. Drives like he's ticked-off at what he's doing because he always wanted to test rocket sleds in the desert.

Certainly the above are highly extreme examples for each category. They are meant to amuse you but also cause you to think. If, for some strange reason, one of your drivers approximates the attributes above, this might be a good time to call "Time-Out" and schedule a meeting to discuss things.

VOLUNTEER FIREMEN'S INSURANCE SERVICES, INC.
PERSONNEL FILE

(Attach Photo Here)

Name: _____
      (Last)   (First)   (Middle In.)

Address: _____

City of Town: _____ State: _____ Zip: _____

Telephone #: _____ (H) _____ (B)

Employer: _____

Address: _____

City/Town: _____ State: _____ Zip: _____

Social Security No: _____ Driver License No: _____

Married: _____ Year: _____ Spouse's Name: _____

Beneficiary: 1 st _____ 2nd: _____

Dependents: _____
      Name   D O B   Name   D O B

      Name   D O B   Name   D O B

Date Joined Dept:Aate Terminated: _____ Reason: _____

EQUIPMENT ISSUED

Item	Ser. # or Size	Date Iss.	Date Ret.

OFFICES HELD

Title	From	To	Remarks	By

52

"TO REVIEW FACTORS DISCUSSED"

HUMAN ASPECTS

1. Attitude

2. Knowledge

3. Judgment

4. Mental Fitness

5. Habits

6. Driving Characteristics

7. Physical Fitness

8. Age

ACQUIRED ABILITY

1. License

2. State & Local Laws

3. Vehicle Characteristics

   Λ) Type

   B) Components

      1) Engine

      2) Drive Train

      3) Miscellaneous

4. Handling Vehicles

SECTION V
LEGAL ASPECTS

LEGAL ASPECTS

Each of us. in the Emergency Service, is accountable to someone due to our actions. Be this entity the Fire Department or the government which approves our organization to provide these services. Our capabilities, or lack of same, are evaluated by our actions.

Most fire organizations are legal entities, by definition, which can sue and and be sued. Too easy, in the past, was it to offer haphazard or lax service, falsely believing that the public would be intimidated against filing a lawsuit due to improper or negligent actions.

Some in the Fire Service mistakenly thought their actions were impossible to view--that the public was ignorant to functions performed at an accident; that an unknowing, non-understanding citizen could not be critical of actions taken. Somehow, through bluster and bluff, the public was made to think that the Fire Service was inviolate.

How mistaken, especially in this time of people demanding quality services and filing a lawsuit against substandard results. Anyone can sue another. Exceptions exist in areas of immunity. But just as the immunity adage was; "The King can do no wrong," in support of immunity, this has been changed to; "The King *shall* do no wrong." It's a Magna Carta, of sorts, for the citizen.

How do Fire Organizations become parties (defendants) in these legal entanglements?

Negligence is one aspect. Negligence is defined as; "habitual failure to do the required thing; carelessness in manner or appearance; indifference." 3) in law, "failure to use a reasonable amount of care when such failure results in injury to another." Elementally, we may view the particulars individually to examine the matter:

"Failure to use. . ."	expresses that means were available, or prescribed which should have been used, but were not.
". . .a reasonable amount of care. . ."	what constitutes reasonable amount of care may be the hingepin in negligence cases.
". . .when such failure results in injury. . ."	A negative result is the end-product of the failure; in this case, injury.
". . .to another."	Not to yourself, but to another indicates that the proximate cause of the injury, due to your failure resulted in a negative occurrence to another party.

Lack of acting "in good faith" may be viewed as a contributing factor to negligence. "In good faith" denotes actions were taken, and viewed as appropriate to reduce the chance or likelihood of negative occurrences.

This would be recognizing the duties of the Fire Organization and move to; 1) increase efficiency of personnel, 2) increase effectiveness of the organization, 3) reduce the threat, or 4) reduce loss through pre-fire planning, resource management and training programs.

What are you legally bound to do as an Emergency Service Organization, this entity which can sue and be sued?

What does your charter state you will do? It is a legal document, recognized by the municipality which has granted you permission to act for them. What about your By-Laws and Constitution? You are a legal organization

58

which must comply with rules and regulations of the society in which you function. The Charter does not set you above or beyond normal laws and standards. It does not put you "out of reach."

Just as the Sociologist Wolfe stated: "If we define situations as real, they are real in their consequences." The consequences of thinking you are guarded, by being in the Fire Service, against recrimination or suit, can and has led to a false sense of security.

These mistakes can and have been costly to; 1) the Fire Chief, 2) the Fire Company/Department, and 3) municipality of jurisdiction, to say nothing of individuals named in the suit.

This statement infers that a suit against, for example, the driver of a department vehicle, may, through a ripple effect, draw-in not only the driver, but all parties above him within the authority structure. As the legal pot "boils," there will/may be those who disengage themselves from this net. The mere inference of Fire Companies/Departments and municipalities being named in suits (even if they are ultimately released from the suit) brings a sting of discredit to those named.

The backlash against future occurrences by the organization and/or municipality will place the participants under a microscope. This could make future relationships tense or outright unbearable. In some instances, due to the conditions of the suit and findings of the case, stations have been shut-down, charters dissolved and fire protection contracted from neighboring municipalities. The judgment from a case may force a company into bankruptcy.

For a better understanding of how deeply and tightly this labyrinth may wind does not require a law degree. However, many Fire Service Administrators may never have spoken to the Solicitor or to an attorney to discover if their agency's actions were on firm ground or "running the razor's edge."

And just as the Fire Service is diversified, so is the legal arena. Just because an act or omission may not fall under criminal statutes, does not mean civil law may not apply.

To get you moving toward a better understanding, thereby reinforcing your positive actions and attempting to help you halt or interdict potentially harmful acts, the following texts should be read:

1. *Introduction to Fire Protection Law,* D. L. Rosenbauer, National Fire Protection Association, Boston, MA, 1978.

2. *Legal Insight,* H. Newcomb Morse, 2nd Edition, National Fire Protection Association, Boston, MA, 1975.

3. *Fire-Related Codes, Laws, and Ordinances,* Vince H. Clet, Glencoe Publishing, Encino, CA, 1978.

Certainly these texts are but the tip of the iceberg of what is available. Visit a law library. Personnel there will assist you in researching pertinent cases. See what others did or did not do and the consequences suffered, then take steps within your own organization to move away from the possibility of similar actions happening to you. Certainly, laws vary from State to State, but that which constitutes the actions of a "reasonably prudent individual" bears parallel similarities.

We will concentrate on those legal aspects pertinent to you as a driver of Emergency Vehicles. Topics to be viewed will include: the legal aspects of EV operations, interpreting the law, what constitutes a true emergency, due regard for others, and finally, a view at laws and regulations of the State.

Legal Aspects of Emergency Vehicle Operations

A. Three types of regulations to follow:

    1. motor vehicle and traffic laws enacted by the state government,

    2. Local ordinances, and, 3. departmental policy about what you, as an emergency vehicle operator, may or may not do.

B. Three principles of Emergency Vehicles operation.

    1. Emergency Vehicle operators are subject to all traffic regulations unless a specific exemption is made.

See App. A-2

      * A specific exemption is a statement which appears in the statutes such as: "The driver of an authorized emergency vehicle may exceed the maximum speed limits so long as he does not endanger Life or Property."

    2. Exemptions are legal only in emergency mode.

    3. Even with an exemption, the operator can be found criminally or civilly liable if involved in an accident.

C. Interpreting the Law:

    should events occur requiring an analysis of driver action, by legal authorities

    your actions will be judged by others, from at least two aspects:

    1. True Emergency

    2. Due Regard

What is a True Emergency?

Exemptions would be legal only in a true emergency situation. Often the definition of true emergency is unclear.

A. Sometimes the operator does not have to decide for himself.

1. The Code system used will make the severity of the emergency clear.

2. Information provided by and *solicited from* the dispatcher will make the nature of the emergency clear.

3. The emergency service with which the operator is affiliated will make the nature of the emergency clear.

B. If the operator must decide if a true emergency exists, the following definitions should be considered:

A situation in which there is a *high probability of death or serious injury to an individual,* or *significant property loss,* and *action by an Emergency Vehicle operator may reduce the seriousness of' the situation.*

TRUE EMERGENCY . . .

A situation in which there is a high probability of death or serious injury to an individual, or significant property loss, and action by an Emergency Vehicle operator may reduce the seriousness of the situation

What is due regard for safety of others'? This is based on circumstances.

Guidelines:

A. *"Enough" notice of approach, before a collision is inevitable.*

1. "Enough" is difficult to define. If motorists have windows up, heater or air conditioner and radio on, it may take them a long time to respond.

2. Notice is given by using appropriate signalling equipment in accordance with statutes.

DUE REGARD . . .

'Enough' notice of approach, before a collision is inevitable

B. In judging due regard in use or signalling equipment, courts will consider:

    1. Was it necessary to use the signal?

    2. Was the signal used?

    3. Was signal audible and/or visible to motorists and pedestrians.

C. Use of signalling equipment must be accompanied by caution.

D. An accepted definition of due regard is: "A reasonably careful man; performing similar duties and under similar circumstances, would act in the same manner."

Case History

Instructor should discuss typical emergency calls to see if they do meet the legal requirement of a true emergency.

    Example: Auto Accident (FD)

           Stand By (FD)

           Emergency Transport (Amb.)

           Etc.

Summary :

Guidelines to minimize negligence:

A. True emergency must exist before exercising exemptions.

    1. High probability of death or serious injury.

    2. Property is imperiled.

    3. Action on operator's part could reduce severity.

B. Under any and all circumstances, exercise due regard.

SECTION VI
PHYSICAL FORCES

Physical Forces

Whereas characteristics of drivers and fitness of apparatus are important, physical forces must be studied. These forces are always evident and will affect the manageability of response. These physical forces, although often thought of in terms of mystery, are really quite understandable.

Since such forces may to a greater or lesser degree serve as "green-eyed monsters" with respect to safe driving abilities, each will be addressed individually and then collectively. Although, as stated, ever present during vehicle operation, the utilization of techniques which will be stressed may lessen or reduce their negative effects upon the vehicle.

We will explain each force, how it is created (conditions) and its behavior which can be detrimental to vehicular response, we also will discuss counter measures which you, the EV driver may use, to avoid tragic results.

The physical forces and laws to be dealt with include;

1. Velocity

2. Friction

3. Inertia

4. Momentum

5. Reaction

6. Centrifugal Force

7. Centripetal Force

These physical forces influence the amount of control the EV operator may possess. If the limits created by the physical forces are not exceeded, the operator can fully control the EV's speed and direction. If these boundary limits are exceeded, control will be lost.

Some actions which may result in lost control, are:

1. Driving too fast for weather, road or tire conditions

2. Accelerating too hard

3. Braking inappropriately

4. Changing direction too abruptly

5. Tracking a curve at too high a rate of speed

The key for EV operators is to know the conditions under which these limits are reached and when ability to control the vehicle will be lost. If you anticipate when such conditions may be realized, you will drive to avoid the occurrence. Avoidance response methods are a defensive means required to preclude potential disaster.

From a safe driving standpoint, you will note that these physical forces deal with potential or kinetic energies. The dynamics which are at work will be the result of kinetic energy.

POTENTIAL ENERGY: "Energy that is the result of relative position instead of motion, as in a coiled spring." Energy capability exists but has not been used.

KINETIC ENERGY: "That energy of a body which results from its motion: it is equal to $1/2mv^2$, where m is mass and v is velocity: As opposed to potential energy."

VELOCITY: "A rate of change of position, in relation to time."
Examples of velocity would be:

1. If a person walks at 3.5 feet/second, that person's velocity is 3.5 feet/second.

2. If a vehicle moves at 50 MPH, the vehicle's velocity is 50 MPH.

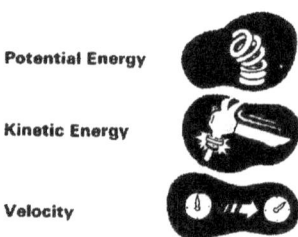

Potential Energy

Kinetic Energy

Velocity

The rates of change of position are measured by the amount of time in which they occur.

While driving, the EV operator can control velocity and direction only.

TYPES OF CONTROL:

1. Velocity Control: This is the control of the EV's rate of speed (motion). This is control over:

    A. Acceleration

    B. Deceleration

    C. Braking

2. Directional Control: This is control of the direction in which the EV will travel. This is typified by:

    A. Steering

    B. Turning

    C. "Tracking" curves in the road

Control of such velocity and direction are of paramount importance and, being the two forces over which the EV driver has control, should be mastered.

It has been stated that an object can have only two types of motion: These two types would be moving with;

1. Constant velocity, or

2. Changing velocity

A changing velocity (increasing) would be termed acceleration.

ACCELERATION: "Change in velocity, either increase (positive acceleration) or decrease (negative acceleration), the rate of such change."

Acceleration is often thought of in terms of increases of velocity (positive acceleration).

If 1 pound is acting in motion, a certain acceleration occurs; if 2 pounds act, the accelerating will be doubled.

Such doubling of weight causing a doubling of acceleration causes a problem when vehicles of greater weight are driven in the same manner that lighter vehicles were operated. Several incidents have occurred where EV drivers operated new equipment in the same fashion as they did lighter vehicles with disastrous consequences.

The most important physical forces for EV control are:

1. Friction

2. Inertia

3. Momentum

4. Centrifugal Force

FRICTION: Friction is defined as "the resistance to motion of two moving objects or surfaces that touch."

From the definition, we may also think of friction as the resistance to slipping. Friction occurs when two surfaces touch or "rub" together.

Points of friction within the EV can be pinpointed for reference:

1. The EV operator's hands on the steering wheel

2. Engine parts rubbing together

3. Gears which mesh

4. Tires and the road surface

5. Brake shoes rubbing on drums/pads rubbing on disc.

From an EV control point, the most important areas of friction are:

1. Friction between tires and road, and

2. Friction between the brakes and the wheels.

TIRE AND ROAD FRICTION: If there were no friction between the tires and the road, the EV would slide all over the highway. We experience this on icy roads. Vehicle control would be impossible. The amount of friction between the tires and road depends upon several things, some of which the EV operator can control:

1. Tire size, tread, type, amount of inflation, etc. (Follow the vehicle manufacturer's specifications with respect to these requirements.)

2. The amount of roiling done by the tires. Friction is:

    A. *Greatest* when the wheels and vehicle are stationary. This seems reasonable since vehicle weight and gravity must be overcome for movement to occur. By standing still, dynamic forces are not at work but static forces are.

    B. *Very Good* - when the wheel is roiling on a dry, smooth road surface. The approximation to standing still may yet be applicable under these conditions but to a lesser degree. Movement itself, begins the introduction of dynamic physical forces to the vehicle. Although changes to the vehicle's behavior may not be readily apparent at low speeds in a straight line, these changes become evident with acceleration, directional change and changing road conditions.

C. *Least* - When the wheel is locked or spinning.

> *Locked:* the normal function of the wheel (to turn) is impeded, within the realm of dynamic forces imposed upon it.

> *Spinning:* the required contact between tire and road surface is broken. Fricitional forces are no longer applied e.g. the two surfaces are no longer "touching" or "rubbing" in the manner needed for controlled contact. One surface (tire) is passing over the other surface (road surface).

FRICTION AT THE BRAKES: This friction is caused by the brake shoes pressing on the drums (or brake pads clamping the brake disc) to create a frictional contact to slow the wheel's turning. Braking or slowing of the vehicle is hereby achieved.

The friction between these two surfaces generates heat. Such may be thought of as the similarity between:

1. Rubbing two sticks together to make a fire.
2. Rubbing your hands together to warm them.
3. Windlassing a rope on a ladder rung to lower a Stoke's litter.

As heat increases between the moving and unmoving surfaces, braking ability decreases. Brake fade is one of the worst consequences of heat due to excessive, hard braking. When sustained (hard) braking sufficiently heats-up the brakes, the pedal-force requirements increase dramatically.

In extreme cases, during such hard brake application, the brakes may suddenly "disappear." The vehicle will continue forward as if no brakes were being applied. At best, it is a scary situation; at worst it may have deadly consequences.

Brake Fade

Brake fade can occur in a variety of ways. In all cases, however, the cause is due to the generation of excessive heat.

1. If the heat generated reaches 700°F or more, the bonding materials of the brake lining melt and act like a lubricant.

2. For some brake lining materials, a gas is generated under high heat conditions. The gas can also act as a lubricant.

3. If the brake fluid becomes too hot, it will expand and reduce braking effectiveness.

4. When the brake lining materials are more than half-gone, the metal frame holding the lining material heats excessively and transfers the heat to the fluid.

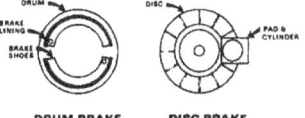

DRUM BRAKE   DISC BRAKE

Drum Brake: Almost 90% of the total drum surface is in contact with the brake shoe at one time. Thus, only about 10% of the surface can be cooling off at one time. (Very little surface area exists to serve as a heat sink).

The brake drums can heat up and expand to the point where it is impossible for the shoes to make good contact with the drums.

Disc Brake: Since the pad makes contact with only about 15% of the disc surface, about 85% of the disc surface is cooling. Disc brake design permits much more cooling. Even if the disc were to get hot, it usually expands and makes *better* contact with the disc pads.

The biggest cause of brake fade in disc brakes is due to worn pads which allow heat to transfer to the hydraulic fluid. Disc pads that are 50% worn have a 300% greater chance of causing fade.

In extreme cases, the heat generated can cause the disc to warp, reduce the amount of contact surface, leading to uneven braking.

VELOCITY CONTROL AND FRICTION

Accelerating

Spinning the wheels reduces friction. Acceleration, then is slowed. Spinning the wheels smooths the tires. The friction between the tires and road surface will be less in the future.

Braking: Shortest stopping distance is achieved by braking so that wheels do not lock up and skid.

Best braking point is just short of locking the wheels.

1. It is difficult to hit exactly that point,

2. Operator may have to jump to jab the brakes.

Locking the Wheels

One of the reasons locked wheels have less friction than rolling wheels is illustrated in the overhead transparency.

Little beads of rubber come off the locked, skidding tires and act like ball bearings on which the vehicle slides.

Changing Direction & Friction

Friction between the tires and road surface is required, if the EV operator is to control the vehicle's direction.

Tires *must* be rolling to change the EV's direction.

1. If brakes lock the front wheels, turning the steering wheel will have no effect on the direction the EV travels.

2. Directional control is possible only *after* brakes are let off and the front wheels begin to roll gain.

As sources of further information, the following pages contain:

1. A chart indicating possible ranges of pavement drag factors for rubber tires.

2. A nomograph for skidmark-speed.

Each of these pages may be used to explore how friction may function on EVs during road and speed conditions.

Since each of the charts deals with friction (skidmarks), an explanation of the Coefficient of Friction is helpful. Remember, friction is described as a force which opposes any force trying to produce motion between two bodies which touch. The ratio of this force (opposition) of friction to the weight of the object is termed the "Coefficient of Friction." It is found expressed as;

$$\text{Coefficient of Friction} = \frac{\text{Force needed to overcome friction}}{\text{weight}}$$

An example is a block of wood weighing 100 Ibs. requiring only 30 Ibs. force to stay in uniform motion:

$$\frac{30}{100} = 0.3$$

If we plotted the "C" for a vehicle, the example may read: A vehicle weighing 30,000 lbs. requires only 15,000 lbs. force to stay in uniform motion:

$$\frac{15,000}{30,000} = 0.5$$

POSSIBLE RANGES OF PAVEMENT DRAG FACTORS FOR RUBBER TIRES

Description of Road Surface	DRY				WET			
	Less Than 30 m.p.h.		More Than 30 m.p.h.		Less Than 30 m.p.h.		More Than 30 m.p.h.	
	From	To	From	To	From	To	From	To
Cement								
New, Sharp	.80	1.00	.70	.85	.50	.80	.40	.75
Travelled	.60	.80	.60	.75	.45	.70	.45	.65
Traffic Polished	.55	.75	.50	.65	.45	.65	.45	.60
Asphalt								
New, Sharp	.80	1.00	.65	.70	.50	.80	.45	.75
Travelled	.60	.80	.55	.70	.45	.70	.40	.65
Traffic Polished	.55	.75	.45	.65	.45	.65	.40	.60
Excess Tar	.50	.60	.35	.60	.30	.60	.25	.55
Brick								
New, Sharp	.75	.95	.60	.85	.50	.75	.45	.70
Traffic Polished	.60	.80	.55	.75	.40	.70	.40	.60
Stone Block								
New, Sharp	.75	1.00	.70	.90	.65	.90	.60	.85
Traffic Polished	.50	.70	.45	.65	.30	.50	.25	.50
Gravel								
Packed, Oiled	.55	.85	.50	.80	.40	.80	.40	.60
Loose	.40	.70	.40	.70	.45	.75	.45	.75
Cinders								
Packed	.50	.70	.50	.70	.65	.75	.65	.75
Rock								
Crushed	.55	.75	.55	.75	.55	.75	.55	.75
Ice								
Smooth	.10	.25	.07	.20	.05	.10	.05	.10
Snow								
Packed	.30	.55	.35	.35	.30	.60	.30	.60
Loose	.10	.25	.10	.20	.30	.60	.30	.60
Metal Grid								
Open	.70	.90	.55	.75	.25	.45	.20	.35

The drag factor or coefficient of friction of a pavement of a given description may vary considerably because quite a variety of road surfaces may be described in the same way and because of some variations due to weight of vehicle, air pressure in tire, tread design, air temperature. speed and some other factors.

These figures represent experiments made by many different people in all parts of the U.S. They are for straight skids on clean surfaces. Speeds referred to are at the beginning of the skid. This table is reproduced from the Accident Investigator's Manual for Police published by the Traffic Institute.
Courtesy, The Traffic Institute
Northwestern University

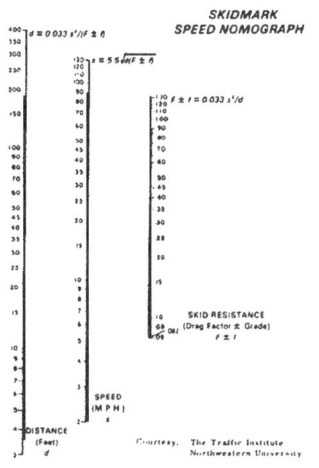

*SKIDMARK
SPEED NOMOGRAPH*

Friction coefficients are also a result of roughness of road surface due to types of materials used. Wet roads reduce frictional contact, therefore reducing the resistance of friction. The vehicle's control, dependent upon such frictional contact, is likewise reduced.

Demonstrate Example using Speed Nomograph

The study of physical forces at work with EVs is best studied by applying those three (3) laws of Sir Isaac Newton (1643-1727). Around 1665, Sir Isaac developed three basic laws regarding the motion of objects. These have come to be known as "Newton's Laws of Motion." These Laws are:

Newton's First Law of Motion: The Law of Inertia

Newton's Second Law of Motion: The Law of Momentum

Newton's Third Law of Motion: The Law of Reaction

1. LAW OF INERTIA: (Also addressed earlier by Galileo, 1564-1642)

"An object continues in its state of rest, or of uniform

motion in a straight line, unless it is acted upon by a net

external force."

It is the natural tendency of an object to maintain its motion. And an object will resist any change in its state of motion.

Inertia is a quality or property possessed by all matter which makes bodies resist any change in their motion. They resist any force that puts them in motion or speeds them up, slows them or stops them once they are moving.

Inertia is inherent in the mass itself. The heavier the mass, the greater the inertial force it exerts against change from a state of rest or change from a velocity.

Inertia has no preference for a heading or direction. It just opposes change.

77

Because of inertia, some outside force must always be supplied to produce or arrest motion. Under ideal conditions (which are rare) without friction, it takes just as much work (energy) to stop a moving body as it does to make the body move.

The amount of force needed to produce this work depends on the distance (kinetic energy) and therefore the time (momentum) during which the force is acting.

1. It takes a much greater force to make a stationary object gain a certain speed quickly than to give it the same speed over a longer period of time.

2. It takes a great amount of force to quickly stop a moving object. A smaller force will stop the object, eventually. But the amount of time in which the desired action is required dictates the amount of force necessary.

3. Vehicular collisions are so dramatic because large objects are stopped in a short period of time. Forces cannot be diminished away from the point of impact, therefore, they are focussed into the impact.

Inertia is also experienced in directional changes of the EV at high speeds. Since the tendency is for an object to move in a straight line, when the vehicle turns to the right, the passengers feel they are moving to the left (actually in a straight line of original travel).

Another example of this "straight line" tendency is viewed in skid marks. Skid marks on a curve are in a straight line, not curved, as some might expect. Again, it is the tendency of the vehicle to continue in a straight line.

Example:
See App. D-2

2. LAW OF MOMENTUM:

"When an unbalanced force acts on an object, the object will be accelerated. The acceleration will vary directly with the applied force and will be in the same direction as the applied force. It will vary inversely with the mass of the object."

The mathematical expression of the Second Law of Newton is:

$$a = \frac{F}{m}$$

More often this Second Law is found written in the following form:

$$F = ma$$

Where: F = Force

m = mass

a = acceleration (velocity)

Simplified, this formula becomes:

$$Momentum = (Mass)(Velocity) = mv$$

Momentum, therefore, is the product of the object's mass (weight) and its velocity.

"The rate of change of momentum is directly proportional to the acting force and takes place in the direction in which the force acts."

A mass of 1 pound, moving with a velocity of 1 foot/second has 1 unit of momentum.

The following may be said concerning the momentum of objects:

1. The momentum of an object has the same direction as its velocity.

2. Whenever the velocity changes for an object, so does its momentum.

3. The mass and velocity of an object will determine its momentum.

Examples:

A. If an Engine and car are driving at the same speed, the Engine, due to its mass (weight), will have a greater momentum.

B. If the car's speed is increased and the Engine's speed is decreased, there would be a point at which the momentum of each would be the same.

C. Consider, however, an Engine with its greater mass (weight) which is also traveling at an excessively high speed. The resultant momentum would be tremendous.

Therefore, to assure a reduction of momentum to safer limits, we have little chance of reducing the mass (weight) of the responding EV once it is on the road. Our next recourse, then, is to reduce the velocity (speed) at which this mass (weight) is moving.

Momentum and Inertia:

As momentum increases, it is harder to overcome the effects of inertia. Momentum and inertia affect velocity control in the following ways:

1. With increased momentum, stopping distance increases.

2. With increased momentum, brakes must work harder, friction and frictional heat must increase.

Directional control is affected by momentum and inertia.

1. With increased momentum, inertia will be harder to overcome. Changing direction, then, is more difficult. Depending upon the amount of momentum, changing direction may be impossible.

2. As momentum increases, the track the EV will follow must be wider.

Law of Conservation of Momentum:

Whenever two objects interact, exactly equal and opposite forces result. The equation of positive forces and negative forces must be balanced. This is stated in the formula;

$$\text{Total Momentum Before Collision} = \text{Total Momentum After Collision}$$

The impulse given by the striking object must be exactly the same as the impulse given to the object struck, but in the opposite direction. A gain in momentum by the struck object occurs through the loss of the same amount of momentum by the striking object.

Therefore; if an Engine, traveling at 50 MPH strikes a stationary Engine of the same weight (mass), the formula for momentum would be;

$$M \times 50\,MPH = 2\,M \times v$$

The Engines would move in the direction of the striking Engine's travel, but with a new velocity of 25 MPH. The struck Engine receives momentum from the one impacting it. The total momentum of both Engines is the same after the impact as it was prior to impact. Now, however, the total momentum after collision is equally divided between the two objects; the one striking and the Engine which is struck.

The formula for momentum of objects colliding which may be of different mass (weight) would be as follows:

$$m_1v_1 + m_2v_2 = m_1v_1 + m_2v_2$$

Where: m_1 = mass of first object

   m_2 = mass of second object

   v_1 = initial velocity of object #1

   v_2 = initial velocity of object #2

   v'_1 = velocity of #1 after collision

   v'_2 = velocity of #2 after collision

3. LAW OF REACTION:

   "For every action there is an equal and opposite reaction."

This describes the action/reaction of forces which occur. In the emergency services, an example evident would be:

REACTION

1. Nozzle reaction from a fire stream. The water issues from the nozzle with a given force, yet a kickback reaction is felt. This is why a flow of over 300 GPM is meant to be applied by a Master Stream device and not by manned lines. The nozzle reaction would be too extreme for control and maneuverability.

CENTRIFUGAL FORCE: "The force, due to inertia, which tends to make a rotating body move away from the center of rotation."

Centrifugal force is caused by the tendency of all bodies to continue to move in the direction in which they began. It has also been termed as the resistance which a moving body offers to deflection from a straight line.

Centrifugal force is that force which tends to push a vehicle, traveling through a curve, away from the center of the turning radius (the inside lane of the curve). This is why race drivers try to head into a curve on the inside, to use as much of the road as possible, realizing this potential for the vehicle to "drift" in a curve.

Examples of Centrifugal Force:

1. Swinging a bucket of water in a wide arc. The water stays in the bucket as it continues to rotate. .

2. High-powered race cars moving to the starting line spin their tires in the "bleach box" for traction. You can see the tires distort and thin as they turn to heat the tread. The tires are moving away from the center (rim) due to this moving away "from the center of rotation."

The Principles of Centrifugal Force:

1. This force is directly proportional to the mass. If we shorten the rope attached to a ball and rotate the ball/rope, the ball is pulled from its path faster than it would be if on a longer rope. The reaction realized with the same weight (mass) ball, but on a shorter rope, is a greater force. The driving application for this is the comparison of forces exerted with gradual directional change as opposed to those forces exerted by making a sharp turn from a forward direction. Sharper turns result in greater Centrifugal Force being exerted when the same speed and weight are used in the comparison.

2. Centrifugal Force is inversely proportional to the radius of curvature, also, as indicated by the shortening of the rope in the above principle.

3. Centrifugal Force is directly proportional to the square of the velocity. If we increase the velocity along the curve, then the pull will be stronger. Expressed;

$$\text{Centrifugal Force} = \frac{m v_2}{r}$$

Where: m = mass in pounds

r = radius in feet

v = velocity in feet per second

g = gravitational constant of 32.2 ft/sec^2

Drivers can feel the centrifugal force when the vehicle negotiates a curve. It is a "push" from the inside of the curve in an outward direction. Centrifugal force is influenced by both speed and the radius of the curve.

1. Sharp turns at high speed = greater centrifugal force.

Centrifugal force is increased four-fold as speed doubles. ("directly proportional to the square of the velocity.")

Example 1: Given a 3,000 pound vehicle entering a 500 feet radius curve:

At 30 MPH: centrifugal force equals 350 pounds approx.

At 60 MPH: centrifugal force equals 1,400 pounds approx.

Example 2: Given a 30,000 pound pumper entering a 500 feet radius curve:

At 30 MPH: centrifugal force equals 3,600 pounds approx.

At 60 MPH: centrifugal force equals 14,400 pounds approx.

2. Tighter curve = greater centrifugal force. As the curve's radius is decreased (the turn becomes sharper), the centrifugal force at a given speed is greater. Thus, a 500 foot curve may be traveled at more than 60 MPH. But control will be lost in a 250 foot curve at less than 50 MPH. ("C.F. is inversely proportional to the radius of curvature.")

AMBULANCE:

$$\frac{mv^2}{r}$$

Mass = 3,000 pounds = $\frac{W}{g}$ =

$\frac{3,000 \text{ pounds}}{32.2 \text{ ft/sec}^2}$ = 93 slugs

Velocity = 60 MPH to ft/sec by; MPH (5280 ft per mile/ 3600 secs per hour) = 88 ft/second.

Radius = 500 feet

C.F. for Ambulance = $\frac{(93 \text{ slugs})(88 \text{ ft/sec})^2}{500 \text{ foot radius}}$ = 1,440 pounds = 1,400 pounds

PUMPER:

Mass = 30,000 pounds = $\frac{W}{g}$ =

$\frac{30,000 \text{ pounds}}{32.2 \text{ ft/sec}^2}$ = 932 slugs

Velocity = 60 MPH to ft/sec by; MPH (5280 ft per mile/ 3600 secs per hour) = 88 ft/second.

Radius = 500 feet

C.F. for Ambulance = $\frac{(932 \text{ slugs})(88 \text{ ft/sec})^2}{500 \text{ foot radius}}$ = 14,434 pounds = 14,000 pounds

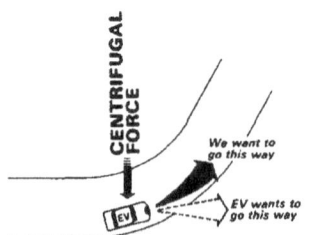

84

CENTRIPETAL FORCE: "Forces proceeding or directing toward the center. The force tending to make rotating bodies move toward the center of rotation."

When a vehicle speeds through a curve, *centrifugal* force acts to carry the vehicle to the outside of the curve. Yet, *centripetal* force is represented by frictional resistance between the tires and road and gravitational force. *Centripetal* force, then, is the resistance offered to the vehicle's movement toward the outside of the curve.

Gravity furnishes centripetal force to objects on the earth's surface. Were it not for the weight of objects, the earth's rapid rotation would permit all bodies on the surface to be projected out, into space in a straight line. There would be an over-abundance of centrifugal force, without the retaining by centripetal force. Again, as we saw a balance or equilibrium concerning pre- and post-collision momentum, this same balance of physical forces may be viewed between *Centrifugal* and *Centripetal* Force.

FOLLOWING ANOTHER VEHICLE

In any given year approximately 150,000 disabling injuries and 500 deaths result from accidents caused by vehicles that were following too closely.* Three things the operator must learn, to be able to follow at appropriate, safe distances:

A. What is a safe following distance?

B. Techniques to help judge or estimate following distance.

C. When to increase following distance.

What is a Safe Following Distance?

An EV operator is following at a safe distance if he can:

A. Stop without mishap if the vehicle in front comes to a sudden stop, or

B. Take evasive action (steer around) to avoid mishap if the vehicle in front comes to a sudden stop.

Estimating Following Distance

* How stopping distance relates to vehicle speed (and weight).

* Relationship between stopping distance and following distance.

* Guidelines to make judgment of the appropriate following distance easier.

A. What is stopping distance?

Reaction Distance + Braking Distance = Stopping Distance

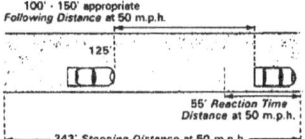

1. *Reaction distance* is the distance a vehicle travels from the time the driver recognizes the need to stop until brake pedal movement begins.

   a. Average drivers require about ¾ second to react.

      * Factors influencing reaction time are:

      * Drivers alertness (fatigue, drugs, allergies, etc.).

      * Driver capability (vision, performance under stress, etc.).

   b. Distance traveled in ¾ second will be greater as vehicle speed is increased.

2. *Braking distance* is the distance traveled from the first brake pedal movement until the vehicle comes to a full stop.

* There is no "average" braking distance. Braking distance varies greatly according to:

* Vehicle speed (higher speed-greater braking distance).

* Vehicle weight (heavier vehicles tend to require greater stopping distances).

* Vehicle condition (brakes, tire tread).

* Road surface, both composition (asphalt, concrete, etc.) and condition (icy, rutted, etc.).

3. *Stopping distances* for various types of vehicles at various speeds are shown in the chart on pages 91 and 92.

   a. All stopping distances on the chart assume driver uses ¾ second to react.

   b. All stopping distances on the chart are based on "hard, dry surfaces."

      * Sedans-about 530 feet

      * Light trucks-about 560 feet

      * Heavy 2-axle-about 610 feet

      * 3-axle-about 680 feet (more than 1/8 mile)

      * The heavier the vehicle, the longer it will take to stop.

      * The higher the speed, the longer it will take to stop.

B. How to tell when the EV is far enough behind.

   1. Following at the full stopping distance (as shown on the chart) is not only unnecessary, it is also impossible!

      * If an EV were traveling that far behind a vehicle in front, other vehicles would constantly pass the EV.

   2. An appropriate following distance will allow enough time to come to a complete stop if the lead vehicle panic stops (stops as fast as possible by braking).

87

* Therefore, safe following distance is greater than the distance required for reaction time, but less than total stopping distance.

MPH (Vehicle speed)
x 2
Minimum following distance (in feet).

3. Two ways to judge following distance:

   a. *Estimate car lengths* - one car length separation is suggested for every 10 mph. A full-sized car is approximately 20 feet long - estimating car length provides *minimum* following distance.

   b. *Two-second rule* - keep a separation of at least two seconds between the EV and the vehicle being followed.

      * Three seconds is a lot safer.

      * Three seconds recommended for larger vehicles.

      * Begin counting (1001, 1002, etc.) when the vehicle in front passes a marker on or beside the road.

      * A pole, sign or tree would be a good marker.

      * Stop counting when the EV passes the same marker. If you passed the marker before the count ended - slow down - you're too close.

      * Car length method - focus of eyes stays constant, but proper estimates are difficult for many people.

      * Two-second method - once learned, allows more precise estimates of adequate following distance.

When Should Following Distance Be Increased?

A. Increase following distance by *50 percent:* if vehicle ahead is unusual, EV is large and/or heavy, EV is not adequately maintained.

   * Fire apparatus would safely use a three-second rule or one *apparatus length* for every 10 mph.

B. *Double* following distance: if road surface is loose or slippery (wet, dirt, gravel), vision is obscured (rain, fog, dust, smog) or driver is not fully alert.

C. *Triple* following distance: if road surface is packed snow or icy.

Practice

A. A driver is returning to the station at the end of an emergency. He is very tired and the road is covered with hard snow. By how much should he increase his following distance?

* The fact that the driver is tired and perhaps not fully alert, would indicate that following distance should be doubled. Since the road is covered with snow, however, following distance should be tripled.

B. A large fire apparatus (elevated platform) is being driven on a high-speed expressway. The operator is taking the apparatus to the city garage for service; some difficulties in the vehicle's braking system have been observed. By how much should the driver increase following distance?

* In this instance, since the vehicle is a large, heavy apparatus, *normal* following distance would be 3 seconds or *1 apparatus length* for every 10 mph. Since the vehicle is not in good condition, following distance should be increased by 50 percent (to approximately 5 seconds or *1-½ apparatus lengths* for each 10 mph).

Consideration: Following Distance in the Emergency Mode

A. In spite of the stress and urgency of an emergency run, the laws of physics do not change. It still takes 243 feet to stop a sedan from 50 mph, and longer for larger vehicles!

B. Should following distance be decreased when traveling in the emergency mode?

* Many operators' reactions and performance get worse under stress. Each operator must learn his own individual capability to respond to stress.

* Motorists may react in crazy ways to lights and sirens. If they stop or slow drastically, the EV operator needs the full amount of following distance to respond.

* A greater following distance permits the EV operator to get the "big picture" of the traffic situation.

STOPPING DISTANCES

Driver reaction distance based on reaction time of 3/4 second, a typical reaction for most drivers under most traffic conditions.

Vehicle braking distance based on provisions of the uniform Vehicle Code for 20 MPH, adjusted where necessary at higher speeds to conform to studies of the U.S. Bureau of Public Roads.

* If perception time is determined to be an appreciable factor for a particular situation, the distance travelled during this time should be added to the total stopping distance. After estimating perception time, the corresponding distance may be determined by reference to the second column which shows the feet travelled per second for the various speeds.

This table developed for educational rather than legal or engineering purposes.

LIGHT 2-AXLE TRUCKS

Speed		Driver # Reaction	Vehicle ## Braking	TOTAL * STOPPING
Miles per Hour	Feet per Second	Distance	Distance	DISTANCE
10	15	11	7	18
15	22	17	17	34
20	29	22	30	52
25	37	28	46	74
30	44	33 11 yds.	67 22 yds.	100 33 yds.
35	51	39	92	131
40	59	44 15 yds.	125 42 yds.	169 56 yds.
45	66	50	165	215
50	73	55 18 yds.	225 75 yds.	280 93 yds.
55	81	61	275	336
60	88	66 22 yds.	360 120 yds.	426 142 yds.

Light 2-Axle trucks

Speed		Driver # Reaction	Vehicle ## Braking	TOTAL * STOPPING
Miles per Hour	Feet per Second	Distance	Distance	DISTANCE
10	15	11	7	18
15	22	17	17	34
20	29	22	30	52
25	37	28	46	74
30	44	33 11 yds.	67 22 yds.	100 33 yds.
35	51	39	92	131
40	59	44 15 yds.	125 42 yds.	169 56 yds.
45	66	50	165	215
50	73	55 18 yds.	225 75 yds.	280 93 yds.
55	81	61	275	336
60	88	66 22 yds.	360 120 yds.	426 142 yds.

HEAVY 2-AXLE TRUCKS

Speed		Driver # Reaction	Vehicle ## Braking	TOTAL * STOPPING
Miles per Hour	Feet per Second	Distance	Distance	DISTANCE
10	15	11	10	21
15	22	17	22	39
20	29	22	40	62
25	37	28	64	92
30	44	33	92	125
35	51	39	125	164
40	59	44	165	209
45	66	50	210	260
50	73	55	255	310
55	81	61	310	371
60	88	66	370	436

Heavy 2-Axle Trucks

Speed		Driver # Reaction	Vehicle ## Braking	TOTAL * STOPPING
Miles per Hour	Feet per Second	Distance	Distance	DISTANCE
10	15	11	10	21
15	22	17	22	39
20	29	22	40	62
25	37	28	64	92
30	44	33	92	125
35	51	39	125	164
40	59	44	165	209
45	66	50	210	260
50	73	55	255	310
55	81	61	310	371
60	88	66	370	436

3-Axle Trucks and Combinations

Speed		Driver #	Vehicle # #	TOTAL *
Miles per Hour	Feet per Second	Reaction Distance	Braking Distance	STOPPING DISTANCE
10	15	11	13	24
15	22	17	29	46
20	29	22	50	72
25	37	28	80	108
30	44	33	115	148
35	51	39	160	199
40	59	44	205	249
45	66	50	260	310
50	73	55	320	375
55	81	61	390	451
60	88	66	465	531

NOTES

3-AXLE TRUCKS AND COMBINATIONS

Speed		Driver #	Vehicle # #	TOTAL *
Miles per Hour	Feet per Second	Reaction Distance	Braking Distance	STOPPING DISTANCE
10	15	11	13	24
15	22	17	29	46
20	29	22	50	72
25	37	28	80	108
30	44	33	115	148
35	51	39	160	199
40	59	44	205	249
45	66	50	260	310
50	73	55	320	375
55	81	61	390	451
60	88	66	465	531

Having viewed the dynamics of physical forces acting upon a vehicle and its occupants, it is proper to now view the benefits of using passenger restraints or seat belts. The need for emergency responders to use seat belts is punctuated by the physical forces just discussed, coupled with the possibilities of being acted upon by the various types of forces throughout the "emergency response corridor" through which your vehicle travels.

US DOT: As emergency responders, we can each recount the tragic results due to not using seat belts. Likewise, the positive outcomes due to passengers using seat belts have been witnessed. The U.S. Department of Transportation, National Highway Safety Administration has made the following observations based on statistical analysis regarding the use of seat belts by the citizenry. This agency projects that each year:

1. 17,000 Lives May Be Saved!

2. The Severity of Almost 4 Million Personal Injuries Would Be Reduced!

3. $20 Billion in Costs Incurred Due to Motor Vehicle Accidents Would Be Reduced!

4. The Total Number of Motor Vehicle Accidents Would Probably Be Reduced!

("The Automobile Safety Belt Fact Book," (DOT HS 802 157), US DOT National Highway Traffic Safety Administration, Washington, D.C. 20590, Rev. 1982.)

The National Highway Traffic Safety Administration, DOT, has structured seat belt requirements for passenger vehicles which may be found in 49 CFR 571.208 Standard No. 208; Occupant Crash Protection. The purpose of this standard is:

"To reduce the number of deaths of vehicle occupants, and the severity of injuries, by specifying vehicle crashworthiness requirements in terms of forces and accelerations measured on anthropomorphic dummies in test crashes, and by specifying equipment requirements for active and passive restraint systems." The dynamic forces viewed earlier are taken into account with respect to vehicle design to ensure a higher level of occupant survival after accidents.

OSHA: Recognizing the need to best ensure driver safety, OSHA, in 1989, proposed "to require driver training programs for all employees who are expected to operate motor vehicles as part of their work assignments. It is OSHA's belief that a properly presented driver training program combined with enforced use of occupant restraint systems can add to the reduction in job-related fatalities."

Section 1915.100 is a new proposed standard and would include the following language related to drivers and seat belt use. The following is excerpted from information provided by the Bureau of National Affairs, Inc., dated September 20, 1989:

Paragraph	*OSHA Is Proposing*
(a)(1)	To cover the safe use and operation of all motor vehicles operated by employees for official business on either public highways or private facilities, on-road and off-road.
(a)(2)(i)	That the regulations would apply to all employees operating motor vehicles as part of their official duties or work assignments.
(a)(3)	The definitions that would be applicable to this regulation. In particular, "Occupant Restraint System" would mean an employee restraint system designed and installed in accordance with 49 CFR 571.208 through .210.
(b)(1)	To require that the employer develope and implement a vehicle safety program that would familiarize all employees with the safe operation of the motor vehicles they would be expected to use as part of their work assignments.
(b)(2)	That, as a minimum, the vehicle safety training program should address the use of occupant restraint systems and driver training.

(d)(1)   That as part of the vehicle safety program proposed in paragraph (b)(1) of this section, the employer must develop and implement an employee driver training program that would train employees, who are expected to operate motor vehicles as part of their official work assignments, in all areas of safe vehicle operation, including the safe use and proper maintenance of occupant restraint systems.

(d)(3)   That employee driver training should be provided by a skilled instructor qualified in the topics being taught.

(d)(4)   That the employer conduct annual employee driver refresher training on the use of occupant restraint systems and safe vehicle operation.

(d)(5)   That all employees hired after the effective date of this section, who are expected to operate a vehicle as a condition of employment, would have to be given the employee driver training proposed in (d)(1) of this section before they would be permitted to operate motor vehicles.

CRASH DYNAMICS:

To better understand the need to use seat belts, crash dynamics should be viewed, specifically focussing on findings of both the US DOT and Canadian government. Studies have shown there are two distinct collisions in an accident; 1) the car's collision, 2) the human collision.

A. *Car's Collision:* Often called the "First Collision," is when a vehicle strikes another vehicle or object.

B. *Human Collision:* Sometimes called the "Second Collision". This occurs when the unbelted occupants are thrown into the vehicle's windshield, steering wheel, doors, dashboard or other hard interior surfaces. This may include person-to-person collisions between occupants or unbelted occupants being ejected from the vehicle compartment and colliding with other objects. The following surfaces have been found to contribute to injuries and fatalities during second collisions.

92.3

Factors Contributing:

Steering Assembly	30%
Exterior of Car and Outside the Car	30%
Surface of Side Interior	22%
Windshield Frame	20%
Roof	12%
Instrument Panel	11%
Windshield	8%
Hood	5%

Unrestrained Occupants:

The following dynamics are based on a test collision wherein two cars, each traveling at 30 MPH crash head-on.

A. *Impact:* Car's Collision (First Collision). The vehicle begins to crush and slow down. Within 1/10th of a second, the vehicle has come to a complete stop. The unbelted occupant, however, continues to move forward in the car at 30 MPH.

B. *Human Collision* (Second Collision). At 1/50th of a second after the car has stopped, the unbelted occupant slams into the windshield. The forces on the occupant in this second collision are equivalent to those on a human body as it hits the ground after falling from a three-story building.

C. *Ejection From Vehicle.* Unbelted occupants may be thrown from the vehicle upon collision or roll-overs. While some may think this is safer than being secured by a seat belt, several factors should be considered. The occupant being ejected from the vehicle's interior will travel a path of many obstructions which may produce injury or death. Such internal obstructions include; other passengers, dashboard, and windshield or door. Once outside the vehicle, the unbelted passenger faces further injury due to whatever may block their path; trees, impacting vehicle and finally the surface upon which they come to rest.

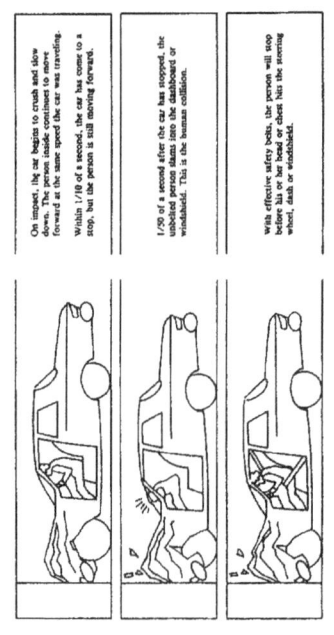

92.4

Belted Occupants:

The following dynamics, provided in the Canadian study, occur with respect to restrained (belted) occupants during collisions:

A. *Impact:* "The car begins to crush and to slow down."

B. *As The Car Slows Down:* "The person moves forward until the seat belts restrain him. The belts keep him in his seat and keep his head and chest from striking the car interior."

C. *Belted Occupants:* "Are able to 'ride down' the collision as part of the car. They are able to take advantage of the car's slower stop, as it crushes and absorbs energy. *For belted people there is no human collision.*"

Myths vs. Facts Regarding Seat Belt Use:

The National Highway Traffic Safety Administration discusses various myths regarding the reluctance of some to use seat belts. These should also be viewed concerning emergency vehicle drivers.

Myth	Fact
A. "I don't need seat belts because I'm a really good driver. I have excellent reactions."	"No matter how good a driver you are, you can't control the other car. When another car comes at your, it may be the result of a mechanical failure and there's no way to protect yourself against someone else's poor judgment and bad driving."

As an emergency vehicle operator, conditions may exist or be created by your presence which may affect other drivers to react in a manner not anticipated. An example would include persons who panic when they hear or see your approach thereby reacting in a manner which causes an accident with your vehicle or another. Defensive driving is the key.

THE MYTHS

The National Highway Traffic Safety Administration discusses various myths regarding the reluctance of some to use seat belts. These should also be viewed concerning emergency vehicle drivers.

Myth	Fact
A. "I don't need seat belts because I'm a really good driver. I have excellent reactions."	"No matter how good a driver you are, you can't control the other car. When another car comes at your, it may be the result of a mechanical failure and there's no way to protect yourself against someone else's poor judgment and bad driving."
B. "I don't want to be trapped in by a seat belt. It's better to be thrown free in an accident."	"Being thrown free is 25 times more dangerous . . . 25 times more lethal. If you're wearing your belt you're far more likely to be conscious after an accident . . . To free yourself and help your passengers."
C. "I just don't believe it will ever happen to me."	"Everyone of us can expect to be a crash once every 10 years for one out of 20 of us, it'll be a serious crash. For one out of every 60 born today, it will be fatal."
D. "Well, I only need to wear them when I have to go on long trips, or at high speeds."	"Eighty (80) percent of deaths and serious injuries occur in cars traveling under 40 miles per hour and 75 percent of deaths or injuries occur less than 25 miles from your home."
E. "I can touch my head to the dashboard when I'm wearing my seat belt so there's no way it can help me in a car accident."	"Safety belts were designed to allow you to move freely in your car. They were also designed with a latching device that locks the safety belt in place if your car should come to a sudden halt. This latching device keeps you from hitting the inside of the car or being ejected. It's there when you need it."
F. "I don't need it. In case of an accident, I can brace myself with my hands."	"At 35 miles per hour, the force of impact on you or your passengers is brutal. There's no way your arms and legs can brace you against that kind of collision. The speed and force are just too great. The force of impact at just 10 MPH is equivalent to the force of catching a 200-pound bag of cement from a 1st-story window."
G. "Most people would be offended if I asked them to put on a seat belt in my car."	"Polls show that the overwhelming majority of passengers would even willingly put their own belts on if only you, the driver, would ask them."

B. "I don't want to be trapped in by a seat belt. It's better to be thrown free in an accident."

"Being thrown free is 25 times more dangerous . . . 25 times more lethal. If you're wearing your belt you're far more likely to be conscious after an accident . . . To free yourself and help your passengers."

C. "I just don't believe it will ever happen to me."

"Everyone of us can expect to be a crash once every 10 years for one out of 20 of us, it'll be a serious crash. For one out of every 60 born today, it will be fatal."

Your emergency response will take you through the same travel corridors as those projected in the statistics. Will your vehicle be a part of those statistics?

D. "Well, I only need to wear them when I have to go on long trips, or at high speeds."

"Eighty (80) percent of deaths and serious injuries occur in cars traveling under 40 miles per hour and 75 percent of deaths or injuries occur less than 25 miles from your home."

"Under 40 miles per hour." Sounds similar to usual response speeds in congested municipalities for emergency vehicles.

"Occur less than 25 miles from your home." What are the distances for your normal response territories, even if mutual aid distances are included?

E. "I can touch my head to the dashboard when I'm wearing my seat belt so there's no way it can help me in a car accident."

"Safety belts were designed to allow you to move freely in your car. They were also designed with a latching device that locks the safety belt in place if your car should come to a sudden halt. This latching device keeps you from hitting the inside of the car or being ejected. It's there when you need it."

To maintain control of the emergency vehicle, the driver must be firmly in place behind the wheel with access to all controls. Seat belts provide this ability of movement yet secure the driver to enable maintenance of control.

92.6

F. "I don't need it. In case of an accident, I can brace myself with my hands."

"At 35 miles per hour, the force of impact on you or your passengers is brutal. There's no way your arms and legs can brace you against that kind of collision. The speed and force are just too great, The force of impact at just 10 MPH is equivalent to the force of catching a 200-pound bag of cement from a lst-story window."

With the responsibilities of providing safe transport for emergency responders, watching traffic, determining what initial and subsequent duties may be required at the scene, the driver of an emergency vehicle has more to consider than how to "brace against a collision."

G. "Most people would be offended if I asked them to put on a seat belt in my car."

"Polls show that the overwhelming majority of passengers would even willingly put their own belts on if only you, the driver, would ask them."

Departmental policies, adopting the intent and concepts of such standards as NFPA 1500 requiring the use of seat belts, will ensure their use.

SECTION VII
VEHICLE MAINTENANCE & RECORDS

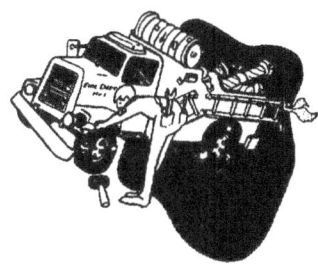

Vehicle Maintenance Records

The maintenance of equipment ensures the safe "use" life of those resources of the organization. By enacting a PM (Preventative Maintenance) program, several objectives are met:

1. The resources are inventoried on a consistent basis ensuring they have not been:

   A. Damaged

   B. Stolen

   C. Lost at Emergency Scenes

   D. Improperly stored against rapid accessibility

   E. Deteriorated to a point of questionable safety

2. Those performing the check continue to be familiar with equipment which, due to infrequent emergency service use, could require retraining in use. Since those responsible for driving/operating vehicles perform this PM, they will retain a working knowledge of the equipment. Also, the addition, modification or deletion of equipment is known by a responsible party. A continuum of awareness is maintained.

3. Obviously these checks discover maintenance needs, for which the checks were intended. However, it also permits a scheduled repurchasing program to be established.

4. Since some jurisdictions, on the County level, create and update Master Resource files for mutual aid and Emergency Management Response, the PM inventory serves to aid this purpose.

5. A documented history of use/distress/testing may also furnish vital input to manufacturers and the Emergency Service using the equipment. Problems discovered, inconsistent with expectations, permit repairs and credits within warranty periods. If drastic conditions are discovered, the manufacturer has an opportunity to publish a recall or stop-use notice. Therefore, such a program does not only benefit your department, but may also help a multitude of agencies,

6. "Readiness" of all factors within an emergency service protection system are provided with a similar "start point." This is to say that all equipment, apparatus, and personnel have systematically been reviewed and evaluated on a scheduled basis and been found to be at a peak performance level. They should function at the highest, acceptable level of safe readiness.

This should be the way each organization meets an emergency:

Apparatus   Functional and mechanically sound. Able to operate over its widest range of capabilities without impediments caused by use or abuse which could have been corrected.

Equipment   As close to purchase conditions as possible. This can be assured through proper use, maintenance and by the performance of acceptable stress and service tests which have been instituted by manufacturers and reputable standards organizations such as the NFPA.

Personnel   Trained on a scheduled basis, as a team, in the use of all departmental equipment. Retraining and advanced courses furnished by recognized groups and instructors on a "competency-based" format.

The stated objectives, deal with the recognition of limitations within the "Emergency Service Protection System." By realizing present conditions of each segment's abilities, a determination can be made of serviceability.

We do as much to assess abilities when we prefire plan tentative responses to potential incidents. Through such an evaluation, we determine our performance level. Example:

> If the fireflow requirement for a given structure is 1,000 GPM minimum and we possess a 500 GPM pumper, we make provisions for additional pieces of apparatus at higher pump capacities. Otherwise our ability to provide the needed service is insufficient.

This same inadequacy exists if we have an abused 1,000 GPM pumper which is leaking at the packing, has not been annually service tested and whose driver/operators have not been trained in the piece's proper use. The capabilities may be far below our expectations. We must meet or exceed the stated needs or our objectives will not be satisfied.

PM precludes the potentially diminished service level by exposing such shortcomings. To have protection reduced due to problems which could have been foreseen is unexcuseable. Although this negligence has been pressed from a legal standpoint, we have the wherewithal] to guard ourselves from such involvement.

The next time you are performing the PM check and feel it is more trouble than it is worth, envision the ramifications to the safety of equipment users without it. Any manner of horrific consequences may occur. It's a parallel to the statement:

"For the want of a nail, the shoe was lost,

For the want of a shoe, the horse was lost,

For the want of a horse, the battle was lost,

For the want of the battle, a kingdom was lost."

From your perspective it might read:

"For the want of PM, the brakes were lost,

For the want of brakes, control was lost,

For want of control, the apparatus was lost,

For the want of apparatus, responders were lost,

For the want of responders, the victims were lost."

The ripple effect caused by uncontrolled physical forces or slack maintenance may be seen.

PM programs document facts, so a false sense of security doesn't prevail with its disasterous consequences. if your personnel are to trust that all is in readiness to protect their lives and those of the citizenry served-prove it! The documents you keep and have available are the record.

There was a form of capital punishment in Europe (England) known as "laying on the stones." The sentenced was laid on his back and a flat wood platform placed over him, exposing the head. The executioners then laid stones on the platform one at a time until the sentenced was crushed. Each seemingly insignificant stone (occurrence) impacting negatively on your department may be thought of as being such a load. Eventually, permitted to go unheeded and unresolved, each could, collectively, crush your organization.

Checklists may be constructed for your own use from manufacturer's manuals after modification. Or as a greater asset, use VFIS Forms in your record system. Its been constructed for clarity and ease of use as well as containing those items which, if properly maintained, will better assure safe use and longevity of your equipment.

SECTION VIII
VEHICLE STANDARD OPERATING PROCEDURES

Vehicle Standard Operating Procedures

This section attempts to dispel the mystery often associated with SOPs (Standard Operating Procedures).

SOPs are not difficult to construct and, in philosophy and use, are not so rigid they are unchangeable. Such changes, however, should be made during the planning process (initial and updating) for maximum effectiveness.

An SOP is a written statement of how an organization will function administratively and operationally. As guessed, it is the transcription, for future reference, of those methods discusssed and agreed upon by Officers and personnel as to how the agency will conduct its business; in-house and during emergencies.

Emergency organizations do as much with recording meeting minutes and other permanent notes. The reason for creating SOPS has several practical purposes. Basically SOPs provide:

1. A lasting record of pre-planned and agreed upon actions.

2. A resource document to train new members and update those initially trained.

3. A means by which required actions have been anticipated well before the actual incident, thereby limiting the time to respond efficiently and effectively to needs.

4. A format to systematically join one set of actions to others, depending upon the incident. This also provides for the ability to expand actions and resources during large-scale operations.

5. When used between multiple agencies of similar emergency services, a standardization is achieved for response. Expectations of response by each agency is assured with reduced on-scene direction.

6 . Better safety is assured to personnel and citizenry due to a commonalty of knowledge as to how multiple agencies will interact.

7. Preservation of equipment is realized due to stating how it will be used under given circumstances.

8. Those actions expected of the individual, the constituent part of the total organization, are stated.

9. A means by which you may show your municipality present and future needs and prove your "professionalism" and dedication to service (in 'Good Faith'!)

SOPs, although the product of forethought, must still be flexible so unforeseen, unique conditions may be dealt with. Naturally, none of us can foresee all eventualities for a given incident type. But, adjustments to response for potential or probable occurrences due to conditions may be accounted for.

Consider how much easier it is to respond to an incident which may require only minor adjustments of performance and resource needs, as opposed to responding "cold". Many emergenices require extensive coordination of manpower and equipment and on the scene is no time to conduct a training program,

Operationally, quite taxing. Consider the difficulty from an administrative standpoint. What if no duty descriptions or task assignments existed for company personnel. There are actually places where the Chief, Assistants, Captains, Lieutenants and Driver/Operators have no job description. Imagine the conflicts which exist in such an agency.

An SOP addresses the following topics, which you have seen used by journalists in every newspaper. These categories clarify:

A. WHO: Indicates which apparatus or personnel perform,

B. WHAT: the needed functions,

C. WHEN: under stated circumstances,

D. WHERE: for a certain incident type,

E. WHY OR HOW: by which methods or evolutions

Often used for Operational SOPs, these questions are answered in the Objectives section prefacing the Operational SOP. The total document addresses each category individually in more depth, An Operational SOP Objective Statement may appear as follows:

WHO: "This SOP will be used by all engine companies and their personnel,

WHAT: to confine or suppress industrial fires,

WHEN: when operating with industrial fire brigades,

WHERE: on plant properties in our jurisdiction,

HOW: utilizing on-site equipment and fire department resources."

The breakdown of these categories parallels the same listing in the "Legal Aspects" section. It is a good way to outline problems and analyze needs.

SOPs should be as concise as practical, yet long enough to reasonably address the topic. Some are very short and miss the mark of answering pertinent questions. Others may be so voluminous, they dwarf the Encyclopedia. The length will be dictated by the complexity of the problem and the means selected to meet the problem as well as anticipate alternatives which may be required.

SOPs should be reviewed and changed as the conditions they address change. Such should be done by a committee chaired by the Chief on a scheduled basis, unless sooner, due to the occurrence of changes.

It may be seen, then, that modifications to the area for which you are responsible will be reflected in your SOPs. This provides you with a way to keep informed as well as to determine if your capabilities to respond are still adequate or if they need refinement.

Provided for you is a copy of the Scarsdale Volunteer Ambulance Corps Driver Manual. (Appendix B) This is a training manual as well as an SOP. It is used to orient drivers with the types and uses of equipment in their charge as well as other tasks expected of drivers.

The document is lengthy due to it dealing with many areas of equipment use and response. You will note that it is an excellent SOP due to establishing a description and location of equipment, how it is to be used and procedures to be followed. The manual also indicates geographic areas which the driver must know to function within the response territory.

Complex? Thorough is a better term. And so are the requirements for those entrusted with the operation of the vehicle. Nothing is slipshod here. This Department has established the standards for its members and supplied a "how-to." The purpose is to train and educate personnel in methods which become reflex actions to best assure safety and efficiency.

Similar SOPs may be written for Engines, Squads, Trucks, Special Units, or Chemical Wagons. They serve as reference documents, training documents and also checklists for officers conducting competency based training on the apparatus. If all which is required by SOP is fulfilled, then the person aspiring for driver status has proven his competence to operate safely. Or at least demonstrated they had knowledge of the required methods.

If the Chief Officer is ever questioned as to what qualifications were required of a person before they were permitted to drive, a record of verification exists.

APPENDIX

APPENDIX A
State Laws/Local Statutes

The following categories are suggested from which your own state's laws may be excerpted and included in this section.

1. Drivers of Emergency Vehicles
2. Reckless Driving
3. Risual Signal on Authorized Vehicles
4. Vehicular Hazard Signal Lamps
5. Passing School Buses
6. Driving or Stopping Close to Fire Apparatus
7. Windshield Obstructions and Wipers
8. Unauthorized Driving Over Fire Hose
9. Obedience to Authorized Persons Directing Traffic
10. Junior Drivers License
11. Local Authorities Liable for Negligence of their Employees
12. Exemption of Entities and Vehicles from Fees
13. Unauthorized Persons and Devices Hanging on Vehicles
14. Exhaust Systems, Mufflers and Noise Control
15. Width of Projecting Loads on Passenger Vehicles
16. Vehicle Entering or Crossing Roadway
17. Duty of Driver on Approach of Emergency Vehicle

MOTOR VEHICLE CODE OF PENNSYLVANIA

"Emergency vehicle." A fire department vehicle, police vehicle, ambulance, blood-delivery vehicle, armed forces emergency vehicle, one private vehicle of a fire or police chief or assistant chief or ambulance corps commander or assistant commander or of a river rescue commander used for answering emergency calls or other vehicle designated by the State Police under section 6106 (relating to designation of emergency vehicles by Pennsylvania State Police).

Section 3105 - Drivers of emergency vehicles.

a. General rule. - The driver of an emergency vehicle, when responding to an emergency call or when in the pursuit of an actual or suspected violator of the law or when responding to but not upon returning from a fire alarm, may exercise the privileges set forth in this section, but subject to the conditions stated in this section.

b. Exercise of special privileges. - The driver of an emergency vehicle may:

   1. Park or stand, irrespective of the provisions of this part.
   2. Proceed past a red signal indication or stop sign, but only after slowing down as may be necessary for safe operations, except as providing in subsection (d).
   3. Exceed the maximum speed limits so long as the driver does not endanger life or property, except as provided in subsection (d).
   4. Disregard regulations governing directions of movement or turning in specified directions.

c. Audible and visual signal required. - The privileges granted in this section to an emergency vehicle shall apply only when the vehicle is making use of an audible signal and visual signals meeting the requirements and standards set forth in regulations adopted by the department, except that an emergency vehicle operated as a police vehicle need not be equipped with or display the visual signals. An ambulance which is transporting a patient may use either the lights or the audible warning system, or both, as determined by the driver of the ambulance.

d. Ambulances and blood delivery vehicles. - The driver of an ambulance or blood-delivery vehicle shall comply with maximum speed limts, red signal indications and stop signs. After ascertaining that the ambulance or blood-delivery vehicle will be given the right-of-way, the driver may proceed through a red signal indication or stop sign.

e. Exercise of care. - This section does not relieve the driver of an emergency vehicle from the duty to drive with due regard for the safety of all persons.
   (Chgd. by L. 1986, Act 51, eff 7/8/86.)

Use of your emergency vehicles requires that you establish standard operating procedures for your organization. VFIS suggests the following as minimum requirements for the safe operation of your emergency vehicles.

(1) The Emergency Vehicle Operators Requirements, listed below, should be adopted by your organization. The document details selection criteria for drivers, training requirements, restrictions on over-age drivers and discipline for drivers with Class A or B violations.

(2) A documented preventative maintenance program should be established. This maintenance program should follow the manufacturer-suggested guidelines. VFIS Maintenance Forms may be used to document your maintenance program.

(3) An Accident Investigation Program should be initiated in your organization. A formal analysis of any vehicle accident or loss should be completed using the VFIS Emergency Vehicle Accident Investigation Form or equivalent. The cause(s) of the loss should be determined and corrective action(s) taken by your organization to try to prevent any future losses.

(Any VFIS forms mentioned above or in the following guidelines may be obtained free from VFIS.)

GENERAL EMERGENCY VEHICLE OPERATORS REQUIREMENTS

VOLUNTEER STAFF (UNDER AGE 21)

1. Due to the lack of general driving experience and considering the amount of training and related activities of a young member of an emergency service organization, all staff under the age of 18 should not be allowed to drive emergency vehicles under *any circumstances*. **NO EXCEPTIONS.**

2. All new candidates to become driver trainees shall be subject to periodic medical evaluation as determined by the governing body of the emergency service organization. The purpose of the physical examination is to determine if the candidate has the physical ability to adequately perform his or her duty as an operator of emergency vehicles.

3. .Between the ages of 18 to 21, any candidate who has demonstrated exceptional abilities with his/her personal driving, may become an emergency vehicle operator trainee. The individual shall remain on this trainee list until his or her 21st birthday. During this time candidates will meet the requirements of a training program established by the local emergency service organization. The training program should include, but not be limited to, the following: Preventive Maintenance, Record Keeping, Legal Requirements, Defensive Driving and Unusual Circumstances Driving. Specific training of vehicle functions - such as Vehicle Systems, Pumps, Tanks, Aerial Devices, Hydraulics, etc. should be included as determined by the emergency service organization. The candidates will also demonstrate their driving ability to the officers of the emergency service organization with the following conditions.

a) The trainee's driving of emergency vehicles shall be limited to training and non-emergency activities. Exception: If an emergency situation occurs and there is not a certified operator present at that time, and the trainee has been trained sufficiently and approved by the officer in charge to act in this capacity, the trainee may operate the vehicle during that emergency.

b) Detailed training records shall be kept on all trainees during their training period. The training records should include, but not be limited to, hands-on experience and classroom time on theory, (four hours classroom/ten hours "hands-on" minimum annually). The VFIS Emergency Vehicle Driver Training Program can be used to complete both hands-on and classroom training. The VFIS record keeping forms also can be used.

c) A Department of Motor Vehicles check be done on every trainee upon initial entry to training phase and annual thereafter until trainee turns 21. This report is to be secured from local sources by the insured. This report should reflect not more than two (2) "Class B" and *no* "Class A" violations in a three year period (see Evaluation Requirements, page A-7).

CAREER STAFF (AGE 18 TO 21)

Since individuals in this age group can be employed as career emergency service personnel and recognizing that their duties would include the operation of emergency vehicles, the following guidelines are established:

a) All new candidates for, or existing operators of, emergency vehicles shall be subject to periodic medical evaluation as determined by the governing body of the emergency organization. The purpose of the physical examination is to determine if the candidate or driver has the physical ability to adequately perform his or her duty as an operator of emergency vehicles.

b) Four hours of classroom training for all new driver training candidates. Periodic classroom training for experienced operators should be performed at the discretion of the chief operating officer.

c) New candidates should have sufficient hands-on training to effectively demonstrate their capability of handling emergency vehicles necessary to perform his/her duties. (10 hrs. minimum)

Experienced drivers should receive annual retraining based upon their actual hands-on emergency vehicle driving activity. The amount of training is to be determined by the chief operating officer.

d) Meet the requirements of a training program established by the local emergency service organization. The training program should include, but not be limited to, the following: Preventive Maintenance, Record Keeping, Legal Requirements, Defensive Driving and Unusual Circumstances Driving. Specific training of vehicle functions - such as Vehicle Systems, Pumps, Tanks, Aerial Devices, Hydraulics, etc. should be included as determined by the emergency service organization.

e) A Department of Motor Vehicles check be done on every trainee upon initial entry to training phase and annual thereafter until trainee turns 21. This report is to be secured from local sources by the insured. This report should reflect not more than two (2) "Class B" and *no* "Class A" violations in a. three year period (see Evaluation Requirements, page A-7).

f) The emergency service organization driver training program and procedure should be based upon current recognized safety standards and policies as well as manufacturer's suggested procedures.

CAREER OR VOLUNTEER FIRE STAFF (AGE 21 TO 65)

1. Drivers should meet the following:

a) All new candidates for, or existing operators of, emergency vehicles shall be subject to periodic medical evaluation as determined by the governing body of the emergency service organization. The purpose of the physical examination is to determine if the candidate or driver has the physical ability to adequately perform his or her duty as an operator of emergency vehicles.

b) Four hours of classroom training for all new driver training candidates. Periodic classroom training for experienced operators should be performed at the discretion of the chief operating officer.

c) New candidates should have sufficient hands-on training to effectively demonstrate their capability of handling emergency vehicles necessary to perform his/her duties (10 hrs. minimum).

Experienced drivers should receive annual retraining based upon their actual hands-on emergency vehicle driving activity. The amount of training is to be determiend by the chief operating officer.

d) All candidates for operators of emergency vehicles shall meet the requirements of a training program established by the local emergency service organization. The training program should include but not be limited to the following: Preventive Maintenance, Record Keeping, Legal Requirements, Defensive Driving and Unusual Circumstance Driving. Specific training of vehicle functions - such as Vehicle Systems, Pumps, Tanks, Aerial Devices, Hydraulics, etc. should be included as determined by the emergency service organization.

e) A Department of Motor Vehicles check be done on each individual every three years. This report is to be secured from local sources by the insured. This report should reflect no more than two "Class B" and *no* "Class A" violations in a three year period. Drivers should voluntarily report any personal violations received (see Evaluation Requirements, page A-7).

f) The emergency service organization driver training program and procedure should be based upon current recognized safety standards and policies as well as manufacturers suggested procedures.

REQUIREMENTS FOR
VEHICLE OPERATORS (AGE 65 AND OLDER)

Drivers over 65 years should not be permitted to drive emergency vehicles in emergency situations. If it is necessary for an individual over 65 years to operate emergency vehicles, the following must be adhered to:

1. Meet all requirements for operators in the 21-year to 65-year class (see Evaluation Requirements, page A-7).

2. An annual physical shall be completed by a licensed physician stating the operator is physically capable of driving an *Emergency Vehicle* in an emergency situation. A signed copy of the completed physical examination must be kept in the member's file. The physical should include, but not be limited to, the following:

   a) No impairment of the use of foot, leg, hand, arm or fingertips, or any other structural defect or limitation likely to interfere with safe driving.

   b) Does not have diabetes mellitus to a degree presently requiring the use of insulin for control.

   c) Has no heart condition likely to cause loss of consciousness or sudden death.

   d) Has no respiratory ailment likely to interfere with safe driving.

   e) Has no arthritic, rheumatic, muscular or vascular condition which interferes with the ability to drive safely.

   f) Does not have epilepsy or any other condition likely to cause sudden loss of consciousness or loss of ability to control a vehicle.

   g) Has no mental, nervous, organic, or functional disease, or any psychiatric condition likely to interfere with safe driving.

   h) Must meet the following minimum vision requirements: At least 20/40 (Snellen) in each eye and in both eyes together, with or without glasses; at least 70 degrees side vision in each eye; the ability to distinguish red, green, and yellow (or amber).

   i) Meet hearing requirements by perceiving a forced whisper at 5 feet with the better ear, or meet specified requirements as measured by a testing device, with or without a hearing aid.

   j) Evaluate medication (if taken) to determine if any chemical impairment would result and interfere with his ability to operate an emergency vehicle.

   k) Must not be diagnosed as an alcoholic.

3. A copy of this Physicians Certificate must be sent to VFIS.

b) Does not have diabetes mellitus to a degree presently requiring the use of insulin for control.

c) Has no heart condition likely to cause loss of consciousness or sudden death.

d) Has no respiratory ailment likely to interfere with safe driving.

e) Has no arthritic, rheumatic, muscular or vascular condition which interferes with the ability to drive safely.

f) Does not have epilepsy or any other condition likely to cause sudden loss of consciousness or loss of ability to control a vehicle.

g) Has no mental, nervous, organic, or functional disease, or any psychiatric condition likely to interfere with safe driving.

h) Must meet the following minimum vision requirements: At least 20/40 (Snellen) in each eye and in both eyes together, with or without glasses; at least 70 degrees side vision in each eye; the ability to distinguish red, green, and yellow (or amber).

i. Meet hearing requirements by perceiving a forced whisper at 5 feet with the better ear, or meet specified requirements as measured by a testing device, with or without a hearing aid.

j. Evaluate medication (if taken) to determine if any chemical impairment would result and interfere with their ability to operate an emergency vehicle.

k. Must not be diagnosed as an alcoholic.

3. A copy of this Physicians Certificate must be sent to V.F.I.S.

   *NOTE: All records are subject to audit by V.F.I.S. or our representatives.

DEPARTMENT OF MOTOR VEHICLES TRANSCRIPT
EVALUATION REQUIREMENTS

CLASS A VIOLATION

An individual who has a Class "A" violation within the past three years normally receives a license suspension from the Department of Motor Vehicles who issued the license. The position of V.F.I.S. with these individuals would be - anyone convicted of a Class "A" violation be suspended from driving our insured's vehicles for a period of 18 months. However, any of these individuals would also be required to attend an approved driver improvement program or equivalent training and be recertified to operate emergency vehicles.

VIOLATIONS

Designation of Type A and Type B violations are based on a survey of state point systems. Violations receiving higher numbers of points are classed as Type A.

Type A Violations

1. Driving while intoxicated.

2. Driving under the influence of drugs.

3. Negligent homicide arising out of the use of a motor vehicle (gross negligence).

4. Operating during a period of suspension or revocation.

5. Using a motor vehicle for the commission of a felony.

6. Aggravated assault with a motor vehicle.

7. Operating a motor vehicle without owner's authority.

8. Permitting an unlicensed person to drive.

9. Reckless driving.

10. Hit and run driving.

Type B Violations

All moving violations not listed as Type A violations. (Exceeding posted speed limit is a Type B violation.)

APPENDIX B

SCARSDALE VOLUNTEER AMBULANCE CORPS

DRIVER MANUAL

A Guide for Training and Operation of SVAC Ambulances

Scarsdale Volunteer Ambulance Corps Driver Manual: A Guide for Training and Operation

1. Driver Responsibilities
2. Driver Qualifications
3. Applicable Sections of the Motor Vehicle
4. SVAC Driving Regulations
5. Operation of the Ambulance When on Call
6. Transport Operation
7. Operation of Vehicle on Emergency Calls
8. Radio Operating Procedure
9. Hospital Routes
10. Locations You Must Know
11. Driver Training Test/Driver Training Instruction (checklist)
12. Driver Training Progress Report

DRIVER RESPONSIBILITIES

The operation of any ambulance carries with it an enormous responsibility on the part of the driver, not only to the patient and the crew traveling in the vehicle, but to the general public, to the drivers of other emergency vehicles, such as police and fire, and a responsibility to protect costly equipment and lives. Any SVAC member, who is presently a driver or who intends to become one must be constantly aware of these responsibilities and must perform accordingly, consciously avoiding unnecessary speed and recklessness, which could lead to serious accidents, injury, and possible death.

The ambulance operator is firstly responsible for the safe transportation of the sick and injured. When transporting the patient to the hospital, it is of utmost importance to remember that the patient is being cared for by skilled crew members and that, except in a few extreme emergency situations, *the ambulance may be operated within normalspeed limits,* thus insuring the safety of the patient and the crew and delivering a smoother ride. The driver must be constantly aware that while he is responsible for the safe operation of the vehicle, the crew chief is responsible for the total operation of the call. The driver must be responsive to the command of the crew chief in regard to how the call is handled.

The second area of driver responsibility includes preventing injury to the general public, bystanders, and pedestrians, as well as automobile occupants, operators, and occupants of other emergency vehicles. The ambulance driver must learn to avoid heavily congested traffic areas and to be careful and alert for pedestrians and bystanders at the scene of an accident. Never take for granted that red lights and a siren will get the ambulance through a congested area. Keep in mind the automobile driver with his windows rolled up, air conditioning and radio on, and his passengers carrying on a conversation. He is not expecting an ambulance, and he may not see or hear it. Be alert for other emergency vehicles, especially when approaching red lights and intersections; a policeman with a siren going and an ambulance driver with a siren going will not hear one another. You must be alert to everything around you. Again, the ambulance driver must be extremely alert to the safety of others.

The third important area of driver responsibility is to the protection of the ambulance and all the equipment it carries, a $40,000 responsibility. The Scarsdale community has generously donated funds to insure that these emergency services are available, and it is the responsibility of the driver to protect that valuable investment with all the skill and care he possesses.

Driver training and driving an emergency vehicle must be approached with seriousness. The need for continued good judgment cannot be stressed enough. It is necessary to be alert, conscientious, and safety-minded at all times with constant regard to protecting the patients, the crew, the public, the vehicle and its equipment. Remember, you are driving a well-known vehicle and if you screw up, the public will remember the name of the rig.

Driver Qualifications

Any active SVAC member, who is at least 21 years of age, who holds a valid New York State driver's license, and who has the consent of the Driver Training Committee, becomes eligible for driver training. The members of the Driver training Committee and crew chiefs will be responsible for the initial introduction to the ambulances and after the introductory session, any qualified WAC driver may assist the trainee with his or her driver education. Driver training will be completed when the trainee has successfully learned to handle the ambulances to the satisfaction of the Driver Training Committee and has passed a practical and written examination and has become familiar with routes to local hospitals. Once the driver training is satisfactorily completed, and only then, will a member become a qualified driver, and with the assignment of the crew chief on call, will be allowed to drive the vehicle on aided calls.

APPLICABLE SECTIONS OF THE MOTOR VEHICLE AND TRAFFIC LAWS

The following are sections taken from the New York State Motor Vehicle and Traffic Law manual, which apply to the operation of emergency vehicles. Every qualified driver and driver trainee is responsible for knowing this information.

s 100-a.   Ambulance. Every motor vehicle designed, appropriately equipped and used for the purpose of carrying sick or injured.

s 101.   Authorized emergency vehicle. Every ambulance, policy vehicle, fire vehicle, and civil defense emergency vehicle.

s 114-b.   Emergency operations. The operation or parking, of an authorized emergency vehicle, when such vehicle is engaged in transporting a sick or injured person, pursuing an actual or suspected violator of the law, or responding to, or working or assisting at the scene of an accident, disaster, police call, alarm of fire or other emergency. *Emergency operations shall not include returning from such service.*

s 375-41.   Colored and flashing lights. The provisions of this subdivision shall govern the affixing and display of lights on vehicles, other than those lights required by law.

1. No light, other than a white light, and no revolving, rotating, flashing, oscillating or constantly moving white light shall be affixed to, or displayed on any vehicle except as prescribed herein.

2. Red lights and certain white lights. One or more red or combination red and white lights, or one white light which must be a revolving, rotating, flashing, oscillating or constantly moving light, may be affixed to an authorized emergency vehicle, and such lights may be displayed on an authorized emergency vehicle when such vehicle is engaged in an emergency operation, and upon a fire vehicle while returning from an alarm or other emergency.

3. Green light. One green light may be affixed to any motor vehicle owned by a member of a volunteer ambulance service, or on a motor vehicle owned by a member of such person's family, or by a business enterprise in which such person has a proprietary interest or by which he is employed, provided such member has been authorized in writing to so affix a green light by the chief officer of such service as designated by the members thereof. Such green light may be displayed by such member of a volunteer ambulance service only when engaged in an emergency operation.

As used in this paragraph, volunteer ambulance service means: a) a non-profit membership corporation (other than a fire corporation) incorporated under or subject to the provisions of the membership corporations law, or any other law, operating its ambulance or ambulances on a non-profit basis for the convenience of the members thereof and their families or of the community or under a contract with a country, city, town or village pursuant to section one hundred twenty-two-b of the general municipal law; orb) an unincorporated association of persons operating its ambulance or ambulances on a non-profit basis for the convenience of the members and their families or of the community.

s 1104.   Authorized emergency vehicles.
a. The driver of an emergency vehicle, when involved in an emergency operation, may exercise the privileges set forth in this section, but subject to the conditions herein stated.
b. The driver of an authorized emergency vehicle may:
1. Stop, stand or park irrespective of the provisions of this title;
2. Proceed past a steady red signal, a flashing red signal or a stop sign, but only after slowing down as may be necessary for safe operation;
3. Exceed the maximum speed limits so long as he does not endanger life or property;
4. Disregard regulations governing directions of movement or turning in specified directions.
c. Except for an authorized emergency vehicle operated as a police vehicle, the exemptions herein granted to an authorized emergency vehicle shall apply *only when audible signals are sounded* from any said vehicle while in motion by bell, horn, siren, electronic device or exhaust whistle as may be reasonably necessary, and *when the vehicle is equipped with at least one lighted lamp* so that from any direction, under normal atmospheric conditions from a distance of five hundred feet from such vehicle, at least one red light will be displayed and visible.
d. An authorized emergency vehicle operated as a police, sheriff or deputy sheriff vehicle may exceed the maximum speed limits for the purpose of calibrating such vehicle's speedometer.
e. The foregoing provisions shall not relieve the driver of an authorized emergency vehicle from the duty to drive with due regard for the safety of all persons, nor shall such provisions protect the driver from the consequences of his reckless disregard fro the safety of others.

s 1144.   Operation of vehicles on approach of authorized emergency vehicle.
a. Upon the immediate approach of an authorized emergency vehicle equipped with at least one lighted lamp exhibiting red light visible under normal atmospheric conditions from a distance of five-hundred feet to the front of such vehicle other than a police vehicle when operated as an authorized emergency vehicle, and when audible signals are sounded from any said vehicle by siren, exhaust, whistle, or bell; the driver of every other vehicle shall yield the right of way and shall immediately drive to a position parallel to and close as possible to, the righthand edge or curb of the roadway, or to either edge of a intersection, and *shall stop* and remain in such position until the authorized emergency vehicle has passed, unless otherwise directed by a police officer.
b. This section shall not operate to relieve the driver of an authorized emergency vehicle from the duty to drive with reasonable care for all persons using the highway.

SVAC DRIVING REGULATIONS

Before any member may begin driver training, he or she must be fully familiar with the location of all standard controls and equipment commonly found on any vehicle plus the locution, function. and use of all special emergency vehicle controls. Drivers will be responsible for knowing the location, function, and operution of all of the following items:

I. *Driving Controls*
1. *Battery switches:* Located under driver seat. The ambulance is always operated with both batteries "on." It is important to remember that the batteries must be turned on before you can start the engine, use the sunction unit, or any lights, and that the batteries should *not* be shut off while the engine is still running.
2. *Volt meter:* It indicates the amount of power being delivered to recharge the batteries from the running engine. If should stay in the green area between 12 and I3 volts, and, if not. should be reported to the person in charge of vehicle maintenance.
3. *Driving Gauges:*
   a. *Oil meter:* Measures the oil level in the engine.
   b. *Fuel gauge:* Measures the amount of gas in the gas tanks. The running tank is selected by a switch labeled fuel located on the bottom of the dash directly beneath the fuel gauge. The gauge always displays the gas level in the operating tank. Each tank should be run until the gauge reads empty before changing the fuel switch to the other tank. The non-operational tank should be checked by drivers to make sure that there is enough gas to serve as a back-up fuel system.
   c. *Speedometer:* Located in the center of the gauges. Measures traveling speed of the ambulance and the total number of miles driven. The mileage of the ambulance should be recorded on the call sheets and in the driving log.
   d. *Ammeter:* Indicates the amount of power being used in the ambulance. The indicator should stay on the center line.
   e. *Engine temperature:* Indicates the temperature of the engine while it is running and should be checked for overheating.
4. *Headlights:* There are two positions: the lower one for parking lights and dashboard lights, and the upper one for headlights. It also operates the overhead light in the driving compartment and the dimmer switch for the dashboard lights. The button for the high beams is on the floor to the left of the brake.
5. *Ignition switch:* The large square key is the ignition key. Accessory position is to the left, on position to the right, and start to the far right. Be sure the batteries are on before starting and remain on while the engine is running.
6. *Wiper-washer:* There are two wiper speeds: low and high. The washer is activated by pushing the control in and pushing the switch left to the off position when finished.
7. *Transmission selector:* The ambulance should be in *Park* while not running or while running but not moving for a period of time. Gear positions include *Reverse* for backing, *Drive* for normal driving, 2 for driving in second gear, and *1* for driving in first gear. These last two are seldom used.
8. *Direction signals:* Located on left side of steering column. Down position for left turn, up for right turn. Be sure to always use them.
9. *4-way flashers:* Located on right side of steering column labeled "hazard." Pull lever out for flashers. These flashers are never to be used on an emergency call.

10. *Horn:* Located in the center of the steering wheel. The horn operates when the black horn/siren switch located on the light console is in the left down position. When the switch is in the right down position, the mechanical siren operates off the horn.

11. *Emergency brake:* Located beneath the left-hand side of the dashboard. The pedal is pushed down to apply the brake and the "brake release" is pulled out to release the brake. The emergency brake should always be checked before driving the ambulance.

12. *Heater-Air conditioner:* Located in the middle of the dashboard. The top lever controls the temperature setting, cool for air conditioning and warm for heater-defrost. The bottom lever selects the function desired.

a. *Vent.* Circulates air throughout the cab from outside.

b. *A/C.* Air conditioner delivers cold air to the cab through vents on sides of dashboard.

c. *Hi-Lo.* Allows the adjustment of the heat flow automatically.

d. *Heat.* Heater delivers warm air to the cab from under the dashboard.

e. *H/D.* Allows the heater and defroster to function at the same time.

f. *Defrost.* Delivers warm air to windshield to clear fog.

The fan switch at the left of the heater controls the fan which delivers heat, air conditioning, defrost or vent at different volume including low, medium, or high.

II. *Emergency Control Console*

1. *Silent signals:* Located on the left side of the panel

a. *Green light:* Topmost light is used to tell the driver when the crew is ready for the ambulance to proceed, or to indicate to the driver that he should be travelling at a faster speed.

b. *Yellow light:* Middle position used to inform the driver to *slow down*. This is an important light to note since it will signify that the crew in the rear feels the speed is unnecessarily fast and that they and the patients are uncomfortable.

c. *Red light:* Bottom light used only to alert the driver to *stop* at *once* and check the situation in the back. When the driver hears the buzzer from the back activated, he should look at the silent signals.

2. *Open door fight:* Red light located on right of the panel. If it lights up, one of the rig doors is not closed, and the ambulance should not be moved until all doors are secure. This safety light covers all of the rear loading doors and all five of the outside compartment doors. It is driver's responsibility to secure all doors.

3. *Beacons:* Red pilot light on switch indicates that the four red and white revolving lights on the corners of the ambulance are working. These should be used on all emergency calls.

4. *Flashers:* Flat red lights in the grill and four (4) red lights on the ambulance corners to flash in an alternate action. These should be used on all emergency calls.

5. *Strobe lights or Fireballs:* Activates the two strobe lights on the hood of the ambulance. They are quite bright and should only be used when heavy traffic requires greater visibility of the ambulance.

6. *External flood fight switches:* These switches should always be left in the most left-hand down position when lights are not in use.

a. *Left flood:* In far left position and operates the two left-hand corner flood lights.

b. *Rear flood:* The middle switch and operates the rear flood lights.

c. *Right flood:* Far right switch which operates the two right corner lights.

7. *Spot lights:* Located on both sides of the driving compartment just above the dashboard. These lights may be used to check street signs or house numbers or may function as scene lights if needed. Switches for these lights are located just above the black direction control knob on the end of each light.

III. *Radio:* The top unit is a "scanner" which can listen to four (4) different channels and holds a specific station when a message is coming over. The radio operator selects the channels to be monitored by depressing any of the channel buttons or turning the indicator dial labeled NP1, NP2, NP3, NP4. In addition, the "scan" button must be depressed so that the radio will scan the selected channels.

In the non-scan mode, the scan button not depressed, the radio will receive on only the frequency which it is set to transmit on. This channel is selected by turning the center selecting knob. In the non-scan mode only one channel can be received at a time.

By placing the radio in the scan position, any number of stations may be monitored and in addition, the selecting knob will control which channel has "priority" over the other stations. This means we will not miss a message on our "priority" channel even if a message can be received at a time.

The microphone for the radio is located to the right of the unit. It should be held about 4-6 inches from the mouth when speaking into it and *should not be used while the siren is running.* To relay a message over the radio, the button on the microphone must be depressed while you speak and released to receive. The radio operating procedure must be read and the format followed all calls.

IV. *Sirens:* The ambulance is equipped with two types of sirens. First, an electronic siren consisting of two roof-mounted speakers and a control unit mounted in the center ceiling of the driver compartment. Second, a mechanical siren.

1. *Electronic siren operation:* Located on the roof of the cab. This siren can only be operated from the control unit. The on/off switch operation is expected. The siren should be left off when not needed to avoid accidental use. The center switch, or "selector" is used to control which sound the siren will make. There are six positions for this control.

    a. Radio-will broadcast all radio transmissions out through, the exterior speakers. The volume out of the speakers can be varied by adjusting the "gain" control.

    b. PA-will broadcast voice out through the speakers by use of the microphone attached to the right side of the siren box. The volume of the PA is also adjusted with the "gain" control.

    c. Manual-will cause a wail siren to be broadcast but only when the red "siren" button (on the left side of the box) is pushed.

    d. Wail-when knob is placed in this position, a continuous wail will be broadcast.

    e. Yelp-when knob is placed in this position, a continuous yelp tone. This can best be used to attract more attention to the moving ambulance, particularly at crowded intersections.

    f. Hi-Lo-another continuous siren tone.

    All three continuous sirens, the wail, yelp, and hi-lo, can only be stopped by turning the selector back to manual.

OPERATION OF THE AMBULANCE WHEN ON CALL

The ambulances are driven on a non-emergency or non-call basis most frequently for the purpose of driver training and for service to both vehicles. The procedure for removing the ambulance from Headquarters should be done in the following manner:

1. Check rear-view and side-view mirrors
2. Turn on both batteries

3. Ignition on and start
4. Release emergency brake
5. Open door
6. Pull out onto apron, clearing doorway
7. Put door down *and check* to make sure it is down
8. Wait for green light on Heathcote Road and *allow traffic to clear*-both directions
9. Turn on beacons and flashers and pass through intersection
10. Once clear of intersection, turn lights off
Il. Proceed with business, obeying all normal traffic laws operating vehicle as non-emergency vehicle

The procedure of returning to Headquarters should be as follows:

1. Always approach Headquarters on Heathcote Road and pull up in front of driveway. If on any other road, turn up Heathcote, down to Medical Center or to the turnaround, and come back.
2. Open door
3. Push interruptor
4. Wait for blue light flash
5. Enter intersection with lights on, facing drugstore doors
6. Back to apron using side-view mirrors to locate your position
7. When clear of intersection, *turn lights off*
8. Back into building using side-view mirrors
9. Position ambulance next to white line
10. Set gear in Park
I I. Turn ignition off
12. Turn off both batteries
13. Close door
14. Log mileage

Transport Operation

SVAC provides a transport service to and from hospitals, home, nursing homes, and doctors' offices to *residents of Scarsdale* who require such services. Operation of the vehicle on a transport basis is the same as operation in a non-emergency basis-i.e. no lights and no siren. All normal traffic regulations must be obeyed. Other requirements would include:

1. Check crew is secure before leaving Headquarters.
2. Leave Headquarters in same manner as described in non-emergency operation.
3. At Scene:
   a. Position ambulance out of flow or traffic
   b. Position for each of loading stretcher
   c. Leave ignition on
   d. Set emergency brake
4. After loading, check to make sure all doors are closed.
5. Check crew and passenger are secure.
6. If pick-up or discharge of patient is done at hospital, back ambulance to emergency entrance for loading or unloading, leaving room for other ambulances to pull in.
7. Turn ignition and one battery off.
8. Driver is always responsible for securing vehicle if he leaves it.
9. All normal traffic must be obeyed when the vehicle returns to Headquarters.

The major factor in determining whether a call should be handled as an emergency call is whether or not the speed at which you travel is going to make the difference between saving the life or losing the life of the patient. An emergency call is one where shift transportation is an important issue, such as in cardiac arrest, respiratory arrest, dangerous shock, or deadly poisoning. The driver should proceed at reasonable speed using the siren only if necessary.

The following is the procedure to use when responding to an emergency or urgent call:

1 . Check house location on map (Mirrors & seat should already be set, driver should do this when he/she first come on duty.)
2. Turn on both batteries
3. Ignition on and start
4. Release emergency brake
5. Push interruptor
6. Open door
7. Pull out into apron, clearing doorway
8. Put door down, and check it in side-view mirror
9. Check crew is seated before pulling out
10. Turn on lights and siren
11. Wait until all traffic is stopped and blue light is flashing
12. The siren should always be used when you are responding in an emergency situation.
13. As cars move to the right, check oncoming traffic and if clear, pass on left.
14. *Intersections:*
    a. Traffic lights-blow siren well in advance of intersection. Slow down as you approach the intersection and check light and traffic. If green, check traffic flow, watching for turning vehicles, then proceed through the intersection looking both ways. If the light is red, *slop,* blow siren continuously, look both ways, and check that all traffic has come to a stop. Proceed through only when safe.
    b. Other intersections (2-way or 4-way stop)-blow siren well in advance of intersection. Slow down checking flow of traffic. Proceed *only* when all traffic has come to a halt. It is necessary to come to a complete stop before proceeding through the intersection.
15. *Arrival at Scene:* Radio out at scene, *stop* and let crew out to check patient, then position ambulance as follows:
    a. out of flow of traffic
    b. for ease of loading stretcher
    c. in direction of hospital
16. At scene, leave ambulance running, put in Park, and set emergency brake, leaving lights on.
17. Once loaded, check all doors, specifically the back door *(driver should be fast one in),* door light, silent signals, buzzer.
18. Once enroute to hospital, radio police enroute and to what hospital. Do not use siren if not necessary.
19. Do not use siren when approaching hospital property.
20. Turn off lights when on hospital property
21. Radio out at hospital
22. Back ambulance to emergency entrance for unloading, leaving room for other ambulances to pull in. Turn ambulance, and one battery off.

23. Assist with unloading patient and if necessary to leave vehicle, *secure* it before doing so and take the key.
24. Return to Headquarters according to regular traffic laws. Radio back in service.
25. Back into Headquarters, using interruptor.
26. Leave rig and equipment clean and orderly and ready for next use.
27. Log your mileage.

RADIO OPERATING PROCEDURES

The driver is responsible for communicating his movement and whereabouts to the Police department, or for designating some one to do this. Drivers and driver trainees must review the radio communication procedure.

There are five times on a call the driver must use the radio:

1) Scarsdale Car 18 or 19 en route to ————————————
(address)

2) Scarsdale Car 18 or 19 10-3 ————————————
(address)

3) Scarsdale Car 18 or 19 en route to ————————————
(hospital)

4) Scarsdale Car 18 or 19 10-3 ————————————
(hospital)

5) Scarsdale Car 18 or 19 10-8 returning to headquarters.

The following is the *HEAR System Procedure:*

1) Find proper 3-number code (posted in rear by suction unit)
2) Put transmit channel selector on 4!! (In front radio control)
*3) Punch in 1st digit (on touch-tone dialer) until white light comes on. Then punch in next 2 digits for *at least 3 seconds each.*
*4) Wait until white light goes out.
*5) Call hospital *exactly as follows:*
   a) Scarsdale Ambulance 19 to ———————————— EMERGENCY.
   (hospital name)

   b) If no answer, wait 15 seconds or so and start again from Step #3 (above)
   c) If still no answer, give up.
*6) When hospital answers give following info:
   a) Patient's age and *sex-NO NAMES!*
   b) What happened (auto accident, etc.)
   c) Symptoms shown by pt.
   d) Vital signs-Pulse, BP, respirations.
   e) E.T.A.
   f) Attending M.D. if known.
7) *Have this info ready before you call!*

Scarsdale Ambulance also has its own frequency, Channel 2 in both rigs. This channel is to be used for communication between Ambulance headquarters and either rig or between rigs or portable radios.

HOSPITAL ROUTES

The ambulance driver is responsible for knowing the most direct routes to the hospitals most commonly used. This will usually involve knowing more than one route to each hospital.

Following are maps and road directions to White Plains, St. Agnes, New Rochelle, Lawrence, United, and Westchester Hospitals.

Scarsdale Volunteer Ambulance Corps
Recommended Hospital Routes Guide:

White Plains Hospital

Post Road Route

1) Post Road (Route 22) Northbound to Maple Ave.
2) Right on Maple Ave. to Hospital entrance.
3) E.R. Entrance on left just past Davis Ave.

Old Mamaroneck Road Route

1) Old Mamaroneck Road Northbound to Mamaroneck Ave.
2) Continue on Mamaroneck Ave. to Maple Ave.
3) Left on Maple Ave. to Hospital entrance.
4) E.R. Entrance will be on the *Right* just past Longview Ave.

Bronx River Parkway Route

1) Fenimore Ave. Exit. Right on Fenimore to Post Road.
2) Left on Post Road (Continue as above).

Alternate Route:

1) Parkway Northbound to WALWORTH X Exit.
2) Right on exit one block to Walworth Ave.
3) Left on Walworth Ave. which becomes Fisher Ave.
4) Fisher Ave. to S. Lexington Ave.
5) Right on S. Lexington Ave. to Maple Ave.
6) Left on Maple Ave. to Hospital entrance which will be on the left.

St. Agnes Hospital

Parkway Route

1) Hutchinson River Parkway Northbound to North Street (Exit #25-Route 27-towards White Plains)
2) Follow North Street to St. Agnes Hospital.
3) E.R. Entrance will be second entrance on left.

Old Mamaroneck Road Route

1) Old Mamaroneck Road to Ridgeway.
2) Right on Ridgeway to North Street.
3) Left on North Street to Hospital which will be on the left.

Post Road Route

1) Follow Post Road Northbound (Route 22) to Armory Place.
2) Bear right on Armory Place onto Westchester Ave.
3) Follow Westchester Ave. around to North Street (Route 127).
4) Hospital E.R. Entrance will be first entrance on the *right.*

Grasslands (Vossberg Pavillion)

1) Take Bronx River Parkway Northbound to Virginia Road Exit.
2) Left on Virginia Road which becomes Grassland Road (Route 100).
3) Continue past Westchester Comm. College to Grasslands Reservation which is on the right.
4) Make right turn into Grasslands and follow signs to Vossberg Pavillion.

United Hospital

1) Hutchinston River Parkway Northbound to Cross Westchester Expressway. (Interstate 287 East towards Rye, Port Chester or CT)
2) Get off at second Route 1 exit.
3) Continue on Route 1 North to High Street (second light)
4) Left on High Street. Hospital entrance immediately on your right.
5) Follow signs around to E.R. entrance.

New Rochelle Hospital

Weaver Street Route
1) Weaver Street to Quaker Ridge Road.
2) Right on Quaker Ridge Road to the end which is North Ave.
3) Left on North Ave. to Lockwood Ave.
4) Right on Lockwood Ave. to Glover Johnson.
5) Left on Glover Johnson one block to Van Guilder Ave.
6) Left on Van Guilder Ave. to E.R. entrance.

Mamaroneck Road Route
1) Mammaroneck Road to Griffen Ave.
2) Take Griffen Ave. all the way to Weaver Street.
3) Left on Weaver to Quaker Ridge Road
4) (Continue as above)

Wilmont Road Route
1) Wilmont Road South to North Ave.
2) Left on North Ave. (then same as above)

Webster Avenue Alternate Route
1) Take either Bon Air Ave. or Second Rd. (right turns) from North Ave. to Webster Ave.
2) Left on Webster Ave. to Lockwood Ave.
3) Left on Lockwood Ave. to Glover Johnson.
4) Right on Glover Johnson to Van Guilder Ave.
5) Left on Van Guilder Ave. to E.R. entrance.

St. Lawrence Hospital

Parkway Route

1) Take Bronx River Parkway Southbound to Desmond Ave. Exit.
2) Right turn at exit onto Desmond Ave. to Dewitt Ave.
3) Right on Dewitt Ave. to Paxton Ave.
4) Left on Paxton Ave. to Palmer Ave.
5) Right on Palmer to Circle, then around circle to W. Pondfield Rd.
6) Hospital E.R. Entrance on left.

Post Road Route

1) Post Road Southbound (Route 22).
2) Bear Right after Mill Road (at fork in road) onto Main Street.
3) Continue straight on Main Street which becomes Midland Ave.
4) Follow Midland Ave. to Pondfield Ave.
5) Right on Pondfield Ave.
6) Follow Pondfield Rd. under Railroad tracks to Circle.
7) Bear right at circle. Hospital E.R. entrance on left.

Locations That You Must Know

Village:

Firehouse 1-Fenimore & Post Roads
            2-Popham & Post Roads
            3-Crossway Road

Highway Garage-Off Ramsey Road
Dump/Incinerator-Off Secor Road
Village Hall-Crane & Post Roads
Village Pool-Mamaroneck Road
Post Office-Woodland Place, Golden Horseshoe Ctr, (N.R.)
Wayside Cottage-Wayside Land & Post Road
Public Library-Olmstead & Post Roads
Police Headquarters-Fenimore & Post Roads
Nature Center-Mamaroneck Road, past H.R.P.

Schools:

Greenacres-Huntington Road, between Sage Terr. & Putnam Rd
Heathcote-Palmer Ave., North of Five Corners
Fox Meadow-Brewster Road, between Butler & Chesterfield
Edgewood-Edgewood between Roosevelt Pl. & Nelson Rd.
Quaker Ridge-Weaver St., past the H.R.P.
I.H.M.-Boulevard & Post Roads
St. Pius X-Palmer Ave. and Mamaroneck Roads
Jr. High-Mamaroneck Road opposite Leather
Stocking Lane
High School-Post Road, Brewster Road, or
Wayside Lane

Parks:

Aspen Road, between Aspen and Springdale Roads
Wynmor, Secor & Wynmor Roads
Drake Road, Drake Road between Nelson & Ferncliff Roads
Davis Park, Boulevard between Lyons & Bradley
Hyattfield, Boulevard west of Post Road
Berkley Pond, between Tisdale & Taunton Roads
Walking area, off Crnr. Valley & Gorham Roads
Georgefield Park, between Post, Oxford, Greendale Roads
Duck Pond, between Heathcote and Sherbrooke Roads

Sports:

Quaker Ridge Golf course, Griffen Ave.
Fenway C.C., Palmer Ave. at W.P. Border
Brite Ave. Tennis, Brite Ave. between Butler & Chesterfield
Wayside Tennis, Wayside Lane, one block east of Fox Meadow Rd
Saxon Woods Gold-Mamaroneck Rd. & H.R.P. entrance

Churches, Synagogues, Etc.:

SC. Baptist Church, Popham & Chase Roads
Trinity Lutheran Church, Crane & Chase Roads
St. James the Less, Crane Rd. & Church Lane
Friends Meeting House, Popham Rd., below Firehouse
Christ, Scientist, Fox Meadow & Wayside
Church of Latterday Saints, Wayside Opp. High School entrance
Scarsdale Synagogue, Ogden Rd. & B.R.P. (can be entered from Parkway exit at Ogden Rd)
SC. Cong. Church, Heathcote & Post Roads
Westchester Reform Temple, Mamaroneck & Myrtle Dale
IHM, Boulevard & Post Roads
St. Pius X, Palmer and Mamaroneck Roads
Hitchcock Presby. Church, Walworth Rd. & Greenacres Ave.
Lutheran Church of the Redeemer, Post & Murray Hill Roads

Other:

American Legion, Mamaroneck Rd. at Scarsdale Pool
Boulder Brook Stable, Mamaroneck & Crossway
Women's Club, Drake Rd. opp. Ferncliff
Ramsey Farm, Ramsey & Palmer
Girl Scout House, Wayside Lane

DRIVER TRAINING TEST/
DRIVER TRAINING INSTRUCTION

Driver Trainee _____
Driver Trainer _____
Date_____

BEFORE LEAVING HEADQUARTERS

_____1. Adjusts mirrors
_____2. Checks gas
_____3. Checks oil
_____4. Checks lights

PULLING OUT OF HEADQUARTERS (check)

_____ 1. Disconnects Life Pak
_____ 2. Batteries on
_____ 3. Ignition on
_____ 4. Release emergency brake (with foot on pedal)
_____ 5. Opens door
_____ 6. Pulls out onto apron, clearing doorway
_____ 7. Puts door down
_____ 8. Waits for green light on Heathcote Road
_____ 9. Waits for traffic to clear
_____ 10. Turns on beacons & flashers
_____ 11. Passes through intersection safely
_____ 12. Turns off lights

RETURNING TO HEADQUARTERS (check)

_____ 1. Always approach Headquarters heading south on Heathcote Rd.
_____ 2. Stop at light
_____ 3. Open door
_____ 4. Wait for green light & allow all traffic to clear front & rear
_____ 5. Turn on beacons & flashers
_____ 6. Enter intersection headed. towards Drug store
_____ 7. Stop in middle of intersection, back up (using side mirrors) into HQ
_____ 8. Position ambulance for free movement around it.
_____ 9. Set gear in park & set emergency brake
_____ 10. Turn off ignition
_____ 11. Turn off both batteries
_____ 12. Close doors
_____ 13. Re-attach Life Pak
_____ 14. Records gas and mileage

HANDLING VEHICLE IN TRAFFIC (Comment)

1. Looks ahead _____
2. Drives within speed limit _____
3. Does not tailgate _____
4. Obeys all traffic signs _____
5. Is alert to other drivers _____
6. Is alert to pedestrians & cyclists _____
7. Yields right-of-way _____
8. Slows down in bumpy areas and pot holes _____
9. Two hands on wheel and drive with one foot _____
10. Moves away from sewer holes _____
11. Drives with window open _____
12. Corners slow & smooth _____
13. Brakes and accelerates slow and smooth _____
14. Signals all moves _____
15. Never assumes right-of-way _____

VEHICLE MANEUVERS (comment)

1. Quick turns right and left ——————————————————————————————————————

2. Tight circle-right ————————————————————————— left————————————

3. Figure 8-right ——————————————————————— left————————————

4. Backing-using side mirrors ————————————————————————————————

   in a circle———————————————————(both ways) into parking place —————————

   stopping when backing———————————————around "S" turns ————————

   straight and steady distance———————————————————————————

5. Stopping in specific spot (with and without skidding) ———————————————

6. Driving up on sidewalks/leaving sidewalks ———————————————————

ADDITIONAL COMMENTS ON BACK

B-16

DRIVER TRAINING PROGRESS REPORT

TRAINEE: _____   TRAINER: _____

Is this your first time out with this driver trainee? Yes _____   No_____

If so, please fill out driver training sheet and file it in the correct place.

If this is your second or more time out, please fill out driver training sheet and fill out progress report here.

Comment on progress of driver trainee from your last practice to this practice.

I feel this driver needs a little more work in certain areas:

   1.

   2.

   3.

   4.

and I have notified him of such.

CHECK ONE:

_____ I feel this person would benefit from another driver instructor's training.

_____ I feel this driver trainee is ready for his test.

_____
Signature

B-17

APPENDIX C

VFIS DRIVING COURSE

OUTDOOR TRAINING SESSION

To The Instructor-

We concluded the classroom portion with a detailed explanation of the hands-on driver training course which i's outlined in NFPA-1002. Fire Apparatus Driver/Operator Professional Qualifications. From experience we have found that this can be set up to demonstrate many of the maneuvers that emergency service personnel will be expected to perform in vehicle operations.

Basic Information Needed To Set Up The Course

1. To effectively set up this driver training course and allow enough room for larger vehicles to maneuver, we recommend that a parking lot with a flat paved surface of 300 feet by 300 feet be sought. The dimensional layout of the specific evolutions can be found on Volunteer Firemen's Insurance Services Driver Obstacle Course Outline. These forms may be requested from VFIS as well as driver operator certification forms for all trainees. These are available at no cost by requesting directly from VFIS.
2. We need approximately 100 red traffic safety cones 18 inches high. These are readily available from power companies, gas companies, construction companies, state highway departments or other resources that you may have available in your community.
3. Other items you will need will be stop watches, scoresheets, clip boards, pencils, barricades and tape measures. A supply of carpenter crayons and a ball of string is also necessary.

The whole course is operated by the driver of the emergency vehicle going to the starting position. Once he is ready, a stop watch will be started and the operator will be continually timed until he finishes the course at Position 7. His total time will be added up and this will become his "driving time." This is all recorded in seconds. Example-5 minutes, 22 seconds: (multiply the 5 minutes by 60 seconds, the score for time would be 322 seconds.) Also, the scorekeeper will keep tabs on the number of cones that are struck, hit, brushed or overturned by the operator. These will be recorded accordingly with the percentage of penalty points.

Once the person has completed the course, we add the time and the penalty points and this becomes his basic score. There are many points that can be learned through judging the penalties and time. Emphasis on safety should be made at all times as well as speed. It is apparent the instructor has the option of disqualifying a person where recklessness or carelessness prevails.

It must be emphasized to the student that this is a course to determine skill advantage, not to speed and try to beat another person's score.

Evolution 1

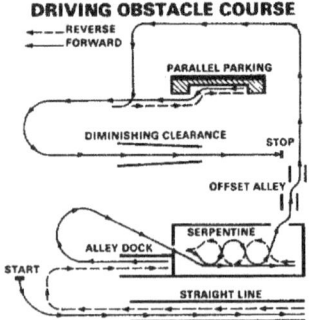

DRIVING OBSTACLE COURSE
- - - - REVERSE
——— FORWARD

PARALLEL PARKING

DIMINISHING CLEARANCE   STOP

OFFSET ALLEY

SERPENTINE

ALLEY DOCK

START

STRAIGHT LINE

Once a driver leaves the starting point, he will enter Evolution 1. Evolution 1 is the straight line exercise. This has a dimension of 8 feet, 6 inches wide by 200 feet long. The object of this particular phase is for the operator to drive from the entrance 200 feet to the end, stop his vehicle, put it in reverse and back out in a straight line 200 feet without hitting any of the cones and also judge the operator's ability to go in a straight line and backward keeping a proper clearance on the left and right side of his vehicle. After he backs out of this Evolution, he will continue to Evolution 2 which is the Alley Dock exercise.

The Alley Dock exercise is designed to test the operators skill in backing into a fire station situation. The dimensions are 10 feet wide by 30 feet deep. We add another dimension to this by putting up a back barricade. This can be accomplished by putting into place two posts and a thin piece of wood or some sort of obstacle to represent a back wall. When the operator backs in and thinks he has the rear portion of his vehicle within 6 inches of this object, he stops. A measurement is taken and appropriate penalty for the measurements is given. (See Score Sheet). Once the operator comes to a complete stop, he immediately starts forward again during which time the measurement will be taken. He proceeds out of the Alley Dock exercise and makes a sweeping wide right hand turn and prepares to enter the Serpentine Exercise.

The driver then drives to the rear of the serpentine exercise area at which time he puts the vehicle into reverse and starts his pattern around the cones as indicated on the score sheet. Once the operator completes the backward section and starts forward to the serpentine, you may start another operator on the course, if you have a second instructor, a second set of stop watches, etc. Again, safety and caution must be exercised at all times.

Driver number one now exits the end or side, depending on the layout of yourcourse, to the Offset Alley. The Offset Alley is an evasive type of maneuver to give the operator the sensation of a sudden change of movement of line change to evade hitting a fixed object. These are 10 feet wide with 34 feet between the corner posts. This judges our operator's ability to make an evasive maneuver. He immediately exits that evolution and goes to probably the most difficult one on the course which is parallel parking. Parallel parking is done blindsided and we make the length of the box 6 feet longer than the particular vehicle the operator is driving. The appropriate penalties are indicated on the score sheet. The operator makes one swing-in to attempt to parallel park. Once he gets the vehicle squared away, he can then proceed immediately out of the parallel parking and go directly to the diminishing clearance.

The Diminishing Clearance Evolution is to give our operator the sensation of driving in a street that starts out at 9 ft. 6 inches and ends up at 8 ft. 2 inches at the end. It is 100 feet long. (Again, this judges his ability to preceive change in dimension.) After this evolution, the operator comes to a stop at our stop sign. There is a measure taken from the front bumper to our finishing barricade and while the operator is returning to the staging area at the starting point, the scorekeeper will then tally all his penalty points as well as his time score and give the operator his completed score. The instructor can note at this particular time if the operator had any particular problems when backing or going forward on the left or right side of the vehicle. For example, if a person hit cones backing on a straight line exercise on the right hand side of the vehicle and also had problems backing through the serpentine judging cones of the right hand side, this can be made apparent to the operator. He then may try to improve his seating position and/or his mirrors on the next run.

Volunteer Firemen's Insurance Services Inc.

DRIVER'S OBSTACLE COURSE

This obstacle course is designed to measure the skills of drivers of emergency vehicles. Through its use, training officials can determine the progress each trainee has made over a given period of time. The "Recommended Time" allocated to each vehicle type indicates an ideal score toward which trainees may use as their objective. It may also be used to test present drivers' skills against a norm. The obstacle course is planned to duplicate seven situations in which driver skill, judgement and knowledge of the limitations of his vehicle are required for effective maneuvering. This course of driving tests is listed in the N.F.PA. Publication #1002 titled FIRE APPARATUS DRIVER/OPERATOR PROFESSIONAL QUALIFICATIONS, 1976, in Appendix A. Scoring is based on total time required to complete the course plus the penalties assigned for mis-maneuvers.

NOTE: CREW MAY ASSIST DRIVER IN ALL OBSTACLES EXCEPT STOP SIGN NO. 7.

NAME: _____

COMPANY: _____

VEHICLE: _____

PENALTY SCHEDULE

OBSTACLE NO.	DESCRIPTION	ERROR	PENALTY
No. 1	Straight Line	Each cone brushed, moved or overturned	10 sec.
		Crossing any line. each time	3 sec.
No. 2	Alley Dock	Each cone brushed, moved or overturned	10 sec.
		Crossing any line, each time	3 sec.
		Stopping 18" or more short of dock stop	10 sec.
		Stopping 12"-17" short of dock stop	6 sec.
		Stopping 6"-11" short of dock stop	3 sec.
No. 3	Serpentine	Each pilon brushed. moved or overturned	10 sec.
		Failure to stop in time, either end of course	10 sec.
		Crossing any line, each time	3 sec.
No. 4	Offset Alley	Each cone brushed, moved or overturned	10 sec.
		Crossing any line. each time	3 sec.
No. 5	Parallel Parking	Each cone brushed, moved or overturned	10 sec.
		Crossing any line, each time	3 sec.
		If distance from curb line is 12" or more	3 sec.
No. 6	Diminishing Clearance	Each cone brushed. moved or overturned	10 sec.
		Crossing any line, each time	3 sec.
No. 7	Stop Sign	Crossing stop line	10 sec.
		Stopping 18" or more short of line	10 sec.
		Stopping 12" to 17" short of line	0 sec.
		Stopping 6" to 11" short of line	3 sec.

SCORE CARD

OBSTACLE NO.:	RUN NO. 1	RUN NO. 2	RUN NO. 3	RUN NO. 4	RUN NO. 5	RUN NO. 6
	Date:	Date:	Date:	Date:	Date:	Date:
1.						
2.						
3.						
4.						
5.						
6.						
7.						
TOTAL PENALTIES +						
DRIVING TIME						
SCORE:						
INITIALS OF SCOREKEEPER:						

DRIVING OBSTACLE COURSE

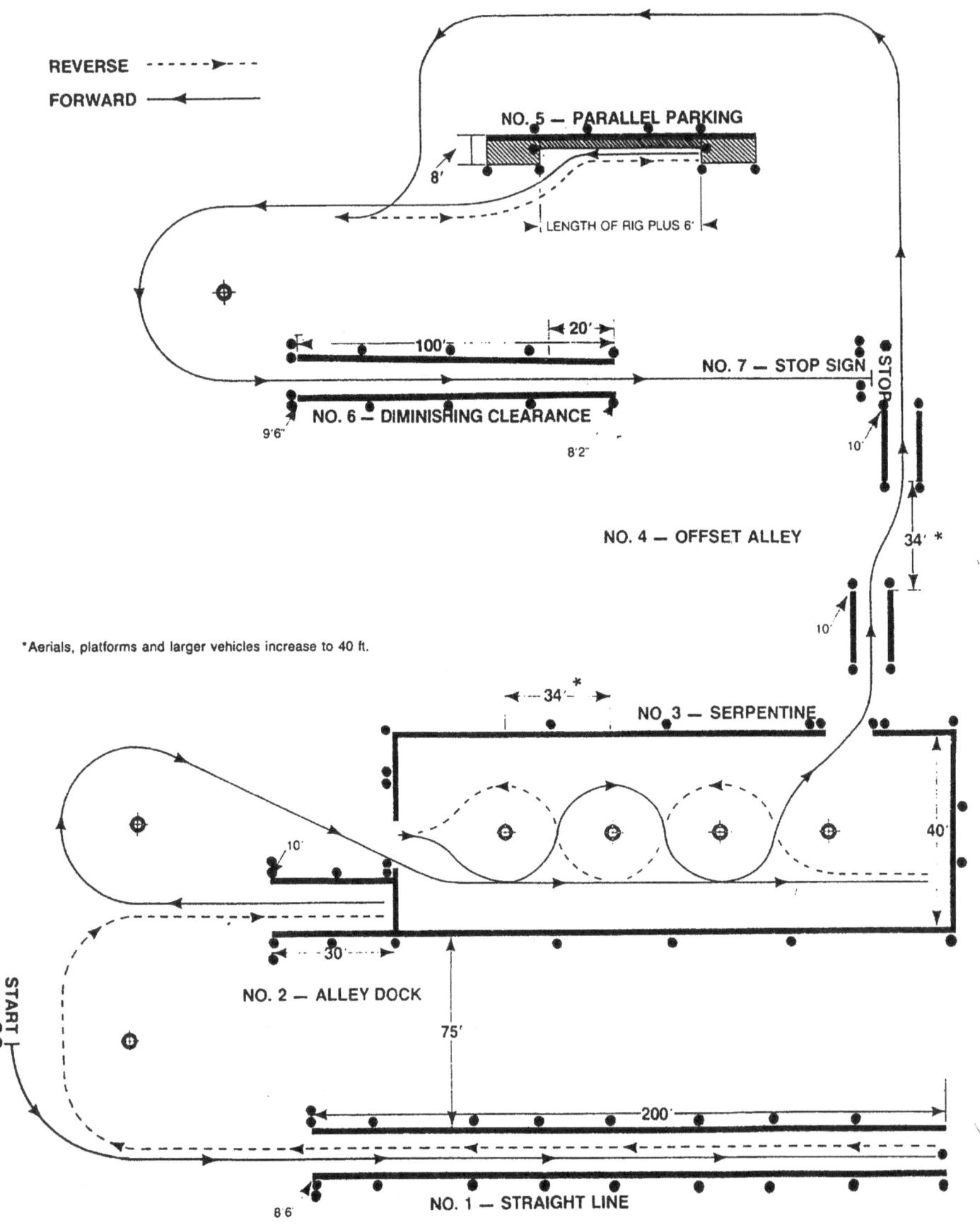

REVERSE ----->
FORWARD ←——

NO. 5 — PARALLEL PARKING

8'

LENGTH OF RIG PLUS 6'

20'

100'

NO. 7 — STOP SIGN

STOP

NO. 6 — DIMINISHING CLEARANCE

9'6"

8'2"

NO. 4 — OFFSET ALLEY

10'

34' *

10'

*Aerials, platforms and larger vehicles increase to 40 ft.

34' *

NO. 3 — SERPENTINE

40'

10'

30'

NO. 2 — ALLEY DOCK

START

75'

200'

NO. 1 — STRAIGHT LINE

8'6"

V.F.I.S. STUDENT WORKBOOK
ANSWER KEY

WORKBOOK SECTION	QUESTION	INSTRUCTORS MANUAL PAGE	LOCATION OF ANSWER
I	A	1	1 st Paragraph
I	B	1	4th Paragraph
I	C	2	Overhead #1
II	A		These answers
II	B		come from general discussion or
II	C		Personal reflection.
II	D thru G	7 thru 26	Student makes notes from 4 of the Scenario's discussed.
II	H	27	V.F.I.S. Stastical Info. Overhead
II	I	28-30	General notes from discussion.
III	A	33-34	Last Paragraph Pg. 33/First Paragraph Pg. 34
III	B	34	Physical Need/Mental Need
III	C	-	General discussion.
IV	A	42	Paragraph 3 & 4
IV	B	43	Paragraph 2
IV	C	44	Section (Acquired Ability)/Items 1,2,3
V	A	58	1st Paragraph
V	B-I	61	B-l
V	B-2	62	1st Sentence
V	B-3	61	B-3
V	C-1	62	Overhead (True Emergency)
V	C-2	62	Overhead (Due Regard)
V	D	61	A
VI	A	68	1 thru 5-top of page
VI	B	69	1 Velocity Control
VI	C	68	Velocity: 1-2 Mi/Hr. Ft./Sec.
VI	D	70	3 of 5 Bottom of Page.
VI	E	72	C
VI	F	72-73	Last two paragraphs/Top of next page.
VI	G	73	Drum & Disc description
VI	H	77-78	Student formulates answers about their response area after understanding Inertia.
VI	I	80	Momentum & Inertia discussion should determine answers.
VI	J	82	#3 Law of Reaction discussion should produce answers.
VI	K-N	82-84	Answers should follow problem demonstration.
VI	O	86	A. What is stopping distance?
VI		88	3 A & B
VI	P	88-90	General Notes
VI	R	92.1-92.7	General Notes
VII	A	95-96	1 thru 6
VII	B	97	General Discussion
VIII	A	103	Definition of S.O.P. 2nd Paragraph & Gen. Discussion
VIII	B	103-104	1 thru 9
VIII	C	104-105	A thru E
VIII	D	105	2nd last paragraph

THE LAW OF MOMENTUM

"When an unbalanced force acts on an object, the object will be accelerated. The acceleration will vary directly with the applied force and will be in the same direction as the applied force. It will vary inversely with the mass of the object."

Formula to express this: $F = ma$

Where: F = Force, in pounds
  m = Mass, in slugs
    a = Acceleration, in feet/second*

Mass is determined by: $m = w/g$

Where: m = mass in slugs
  W = Weight, in pounds
    g = Gravitational constant of 32.2 ft/sec^2

Acceleration is determined by converting MPH (miles per hour) into feet per second2 by:

$$a = (MPH) \frac{5280 ft/mi}{3600 sec/hr} = ft/sec^2$$

EXAMPLE:

"What force is realized by a 30,000 pound engine traveling at 50 MPH?"

$$m = \frac{W}{g} = \frac{30,000}{32.2 \ ft/sec^2} = 932 \text{ slugs}$$

$a = (50 \text{ MPH}) 5280 \text{ ft/mile} = 73 \text{ ft/sec}^2$

$F = ma$
$F = (932 \text{ slugs})(73 \text{ ft/sec})^2$
$F = 68,036 \text{ lbs}$

This formula of $F = ma$ is simplified to determine momentum. It would now be expressed in the following formula:

$$M = mv$$

Where: M = Momentum in lb/sec
  m = mass, in slugs
    v = velocity, in ft/sec.

Mass is obtained for this formula by: $m = \frac{W}{g}$

Velocity is obtained by: $v = (MPH) \frac{5280 \ ft/mile}{3600 \ sec/hr} = ft/sec$

Therefore, using the same 30,000 pound engine at 50 MPH, the momentum realized in pounds/second is:

$$M = mv = \frac{W}{g} v$$

$$M = \frac{30,000}{32.2 \ ft/sec^2} v$$

$M = (932 \text{ slugs}) v$
$M = (932 \text{ slugs})(73 \text{ ft/sec})$
$M = 68,036 \text{ lb/set}$

APPENDIX E
BIBLIOGRAPHY

BIBLIOGRAPHY AND REFERENCES

1. Alabama Law Enforcement Driving Academy. *Curriculum.* Enterprise, Alabama: Author, 1975.

2. American Academy of Orthopedic Surgeons, The Committee on Injuries. *Emergency care and transportation of the sick and injured.* Menasha, Wisconsin: George Banta Co., 1971.

3. The Associated Public-Safety Communications Officers, Inc. *Public safety communications standard operating procedure manual.* New Smyrna Beach, Florida: Author, November 1974 (revised edition).

4. ATCO National Advanced Driver Training and Research Company. *A recommended program for advanced emergency vehicle driver training.* Collingswood, New Jersey: Author.

5. *Automotive Fire Apparatus.,* N.F.P.A. #1901, 1975.

6. Bruch, C.D., *Mechanics for Technology,* New York, New York: John Wiley & Sons, Inc., 1976.

7. California Highway Patrol Academy. *Emergency vehicle operations course in-service driver training* (syllabus of instruction). Sacramento, CA: Author.

8. California Highway Patrol. *Enforcement tactics,* (Chapters 8, 9, 10). California: Author. (HPG 7016).

9. California Highway Patrol, Analysis Section. *Authorized emergency vehicle accidents.* California: Author, December 1974.

10. California State Department of Education. *Driver training* (Fire department driver training manual of instruction, revised edition). Sacramento, California: Office of State Printing, 1972.

11. Center for Safety and Driver Education, Appalachian State University. *Emergency vehicles operations course* (final report). Boone, North Carolina: Author, February 1976.

12. Clark, J.M., Jr. *Emergency and high speed driving techniques.* Houston, Texas: Gulf Publishing Company, 1976.

13. Clet, V.H., *Fire-Related Codes, Laws and Ordinances,* Encino, CA: Glencoe Publishing, 1978.

14. *Code of Federal Regulations,* 49 CFR 571.208, Standard No. 208; Occupant Crash Protection, U.S. Government Printing Office, Washington, D.C. 1987.

15. Council, F.M., Sadof, M.G., Roper, R.B. & Desper, L.P. *Emergency skills resources for range related driver education.* Chapel Hill, North Carolina: University of North Carolina, Highway Safety Research Center, September 1975.

16. Dallas Fire Department, Driver Training School. *Driver training school curriculum.* Dallas, Texas: Author, 1971.

17. Delaware State Fire School. *Driving fire department vehicles.* Delaware: Author.

18. *Elevating Platforms & Aerial Towers,* Fire Officers Guide, P.R. Lyons, N.F.P.A., 1974.

19. *Fire Apparatus Driver/Operator Professional Qualifications,* N.F.P.A. #1002, 1976.

20. *Fire Apparatus Practices,* I.F.S.T.A. 106, 6th Edition, 1980.

21. *Fire Apparatus Maintenance,* 2nd Edition, R. Ely, N.F.P.A., 1975.

22. *Fire Chiefs handbook,* 4th Edition, J. F. Casey, Reuben H. Donnelley Corp., 1978.

23. General Motors Proving Grounds. *Advanced driver education course* (training manual). Milford, Michigan: Author.

24. Greater St. Louis Police Academy,Driver Training Section.*Advanced driver education for emergency vehicle operators.* (Phase I: Classroom; Phase II: Driving Range Accident Avoidance Exercises.) St. Louis, Missouri: Author.

25. Greater St. Louis Police Academy, Driver Training Section. *Advanced driver education for emergency vehicle operators.* (Phase II: Driving Range Accident Avoidance Exercises.) St. Louis, Missouri: Author.

26. *How to set up an emergency driving range.* Boston, Massachusetts: Liberty Mutual Insurance CO. (PR-120).

27. Kansas Highway Safety Coordinating Office, Kansas Highway Patrol Academy: *Emergency vehicle operations course.* Topeka, Kansas: Author.

28. Kent State University. *Emergency driving procedures curriculum guide.* Columbus, Ohio: State of Ohio, Department of Education, Driver Education Section, July 8-19, 1974.

29. Morse, H.N., *Legal Insight,* Boston, MA: National Fire Protection Association, 1975.

30. National Safety Council, Statistics Division, *Accident facts* (1975 edition). Chicago, Illinois: Author, 1975.

31. Naval Safety Center. *Ambulance driver course.* Naval Air Station, Norfolk, VA: Author, 1976. (NAVSAFECEN 1124 OP2).

32. Newcomb, F.D. &Carpenter, K. *Emergency vehicle accident involvement.* New York: State of New York, Department of Motor,Vehicles, Division of Research and Development, 1 July 1972.

33. Oberg, E., and Jones, F.D., *Machinery's Handbook,* 18th Ed., New York, New York: Industrial Press Inc., 1968.

34. *Occupational Safety & Health Reporter,* Published by The Bureau of National Affairs, Inc., Washington, D.C. 20037, September 20, 1989.

35. Office of Emergency Medical Services, Connecticut State Department of Health. *Emergency vehicle drivers training* (course syllabus). Hartford, Connecticut: Author.

36. Oklahoma State University. *International fire service training association manual* (lesson plan no. 15). Stillwater, Oklahoma: Author.

37. Oklahoma State University. *International fire service training association manual* (lesson plan no. 16). Stillwater, Oklahoma: Author.

38. *Operating Fire Department Pumpers,* Fire Officer Guide, P.R. Lyons, 4th Edition, N.F.P.A. 1974.

39. *Operations Aerial Ladders,* Fire Officers Guide, 3rd Edition, N.F.P.A., 1974.

40. Ryder Truck Rental, Inc. *Ryder's fire apparatus driver trainer's program* (syllabus). Doraville, Georgia: Author.

41. Rosenbauer, D.L., *Introduction to Fire Protection Law,* Boston, MA: National Fire Protection Association, 1978.

42. Southeastern State University*Emergency medical technician ambulance* (driver training program). Oklahoma: Author, July 1974.

43. St. Cloud State University.Syllabus-advanced *driving school* (E.M.S.). St. Cloud, Minnesota: Author.

44. St. Cloud State University. *Syllabus-advanced driving school* (fire). St. Cloud, Minnesota: Author.

45. Texas A & M University System, Texas Engineering Extension Service, Fire Protection Training Division. *Emergency driver training in Texas* (diagrams). College Station, Texas: Author.

46. University of North Carolina, Highway Safety Research Center. *Emergency skills resources for beginning drivers.* Chapel Hill, North Carolina: Author, September 1975.

47. U.S. Department of Transportation, National Highway Traffic Safety Administration. *Highway safety plan course.* (administrative and instructional guide). Washington, D.C.: Author, April 1977. (DOT-HS-802-267).

48. U.S. Department of Transportation, National Highway Traffic Safety Administration. *Highway safety plan course.* (instructor materials). Washington, D.C.: Author, 1977. (DOT-HS-802-269).

49. U.S. Department of Transportation, National Highway Traffic Safety Administration. *Highway safety plan course.* (participant materials). Washington, D.C.: Author, April 1977. (DOT-HS-802-268).

50. U.S. Department of Transportation, National Highway Traffic Safety Administration. *School bus driver instructional program.* (trainee study guide). Washington, D.C.: Author, 1974. (DOT-HS-801-087).

51. U.S. Department of Transportation, National Highway Traffic Safety Administration. *School bus driver instructionalprogram.* (trainee study guide-advanced unit). Washington, D.C. Author, 1974. (DOT-HS-801-088).

52. U.S. Department of Transportation, National Highway Traffic Safety Administration. *"The Automobile Safety Belt Fact Book,"* Washington, D.C.: Author, May, 1982. (DOT-HS-802-157).

53. U.S. Department of Transportation, *"The Human Collision, "* Washington, D.C.: Author, reprint of 1st edition prepared, published and distributed by the Ontario Ministry of Transportation and Communications in 1975. (TP-454 CTS-2-76).

54. U.S. Department of Transportation, Transportation Systems Center. *Effectiveness of Audible Warning Devices on Emergency Vehicle* (final report). Washington, D.C. NTIS, 1977. (DOT-TCS-OST-77-38) (DOT-HS-021-948).

55. Virginia Commonwealth University, Highway Safety Training Center. *Virginia emergency operator's curriculum guide.* Richmond, Virginia: Publications Office, Virginia Highway Safety Division, 1975.

56. Walker, L.E. *Emergency vehicle operations course.* Anne Arundel County, Maryland: Anne Arundel County Police Department, 1974.

Department of Transportation
Emergency Vehicle Drivers Course.

Superintendent of Documents
Government Printing Office
Washington, D.C. 20402

Catalogue # 050-003-00332-8
050-003-00330-1
050-003-00331-0

OVERHEAD TRANSPARENCY TEMPLATES

- Personnel Injury or Death

- Peripheral Injury or Death

- Equipment Loss

- Long Term Impact

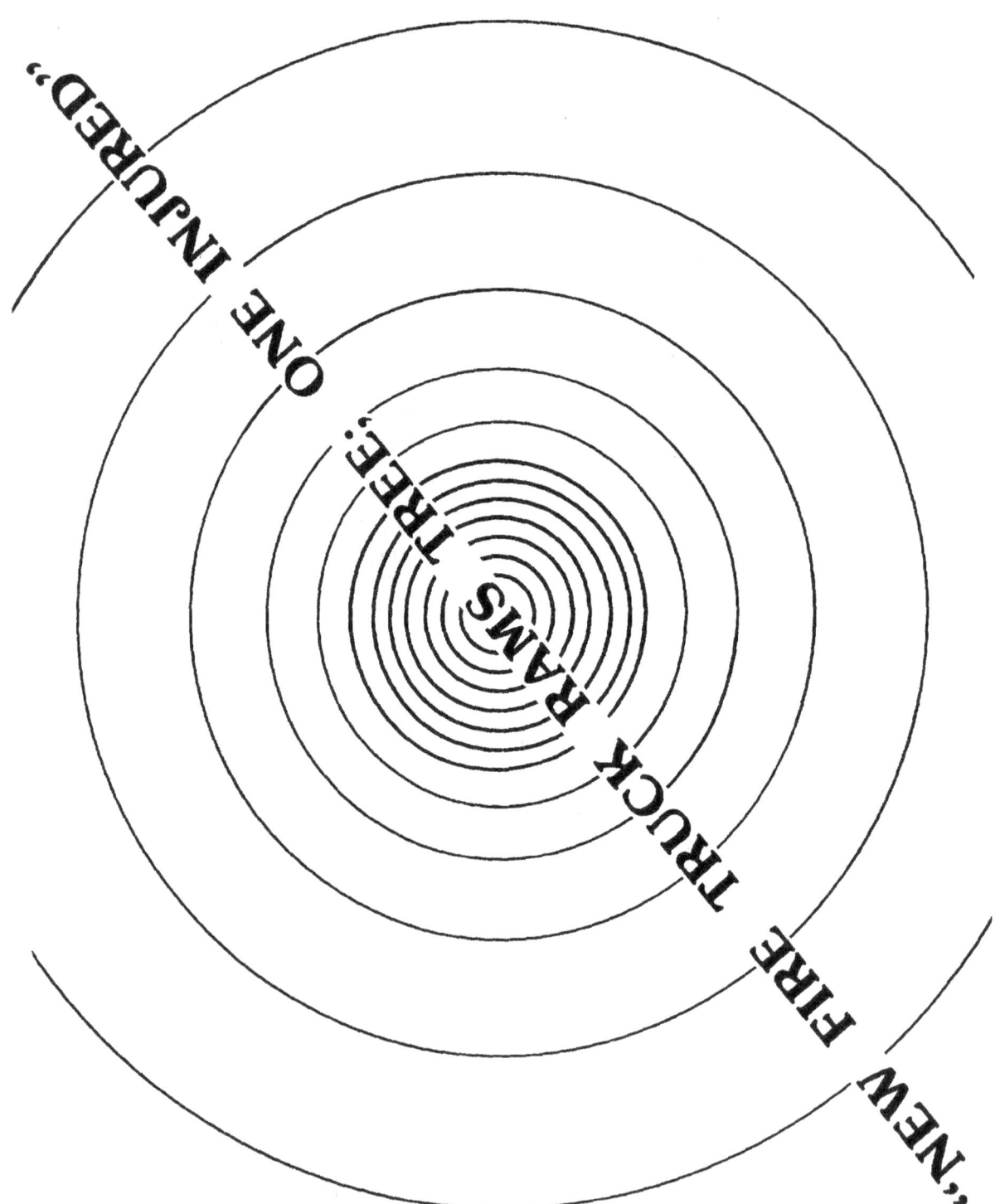

"NEW FIRE TRUCK RAMS TREE; ONE INJURED."

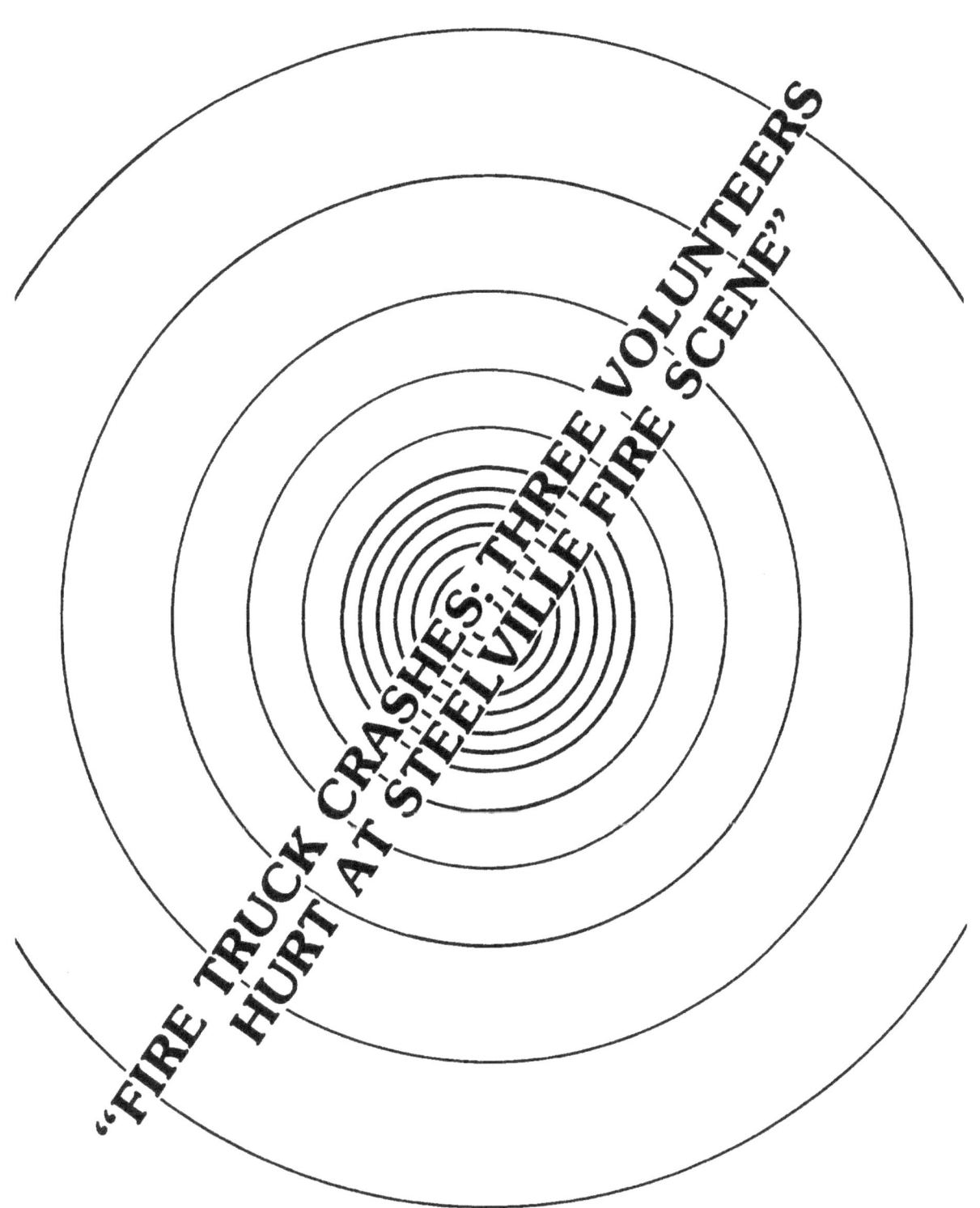

"FIRE TRUCK CRASHES: THREE VOLUNTEERS HURT AT STEELVILLE FIRE SCENE"

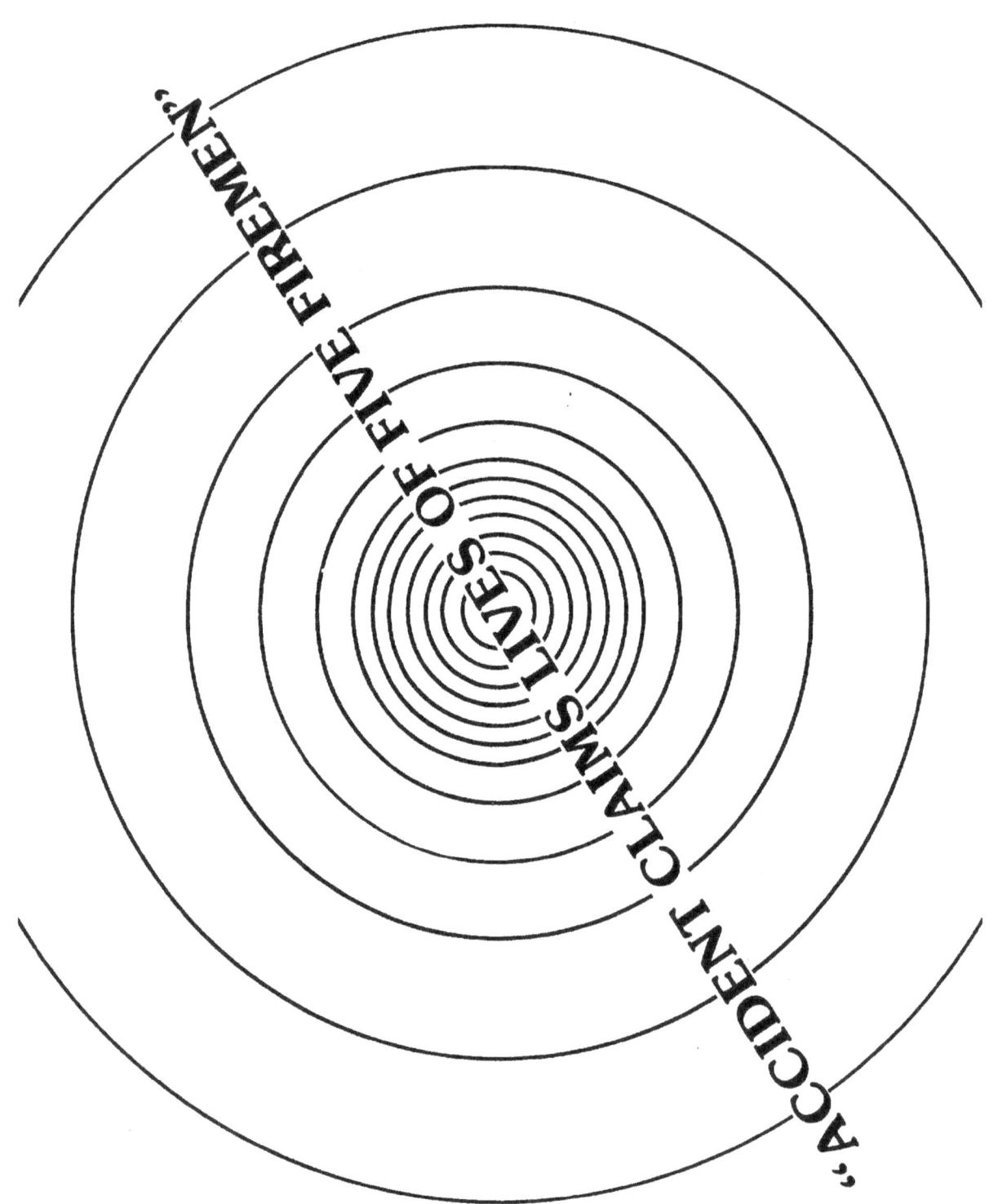

"ACCIDENT CLAIMS LIVES OF FIVE FIREMEN."

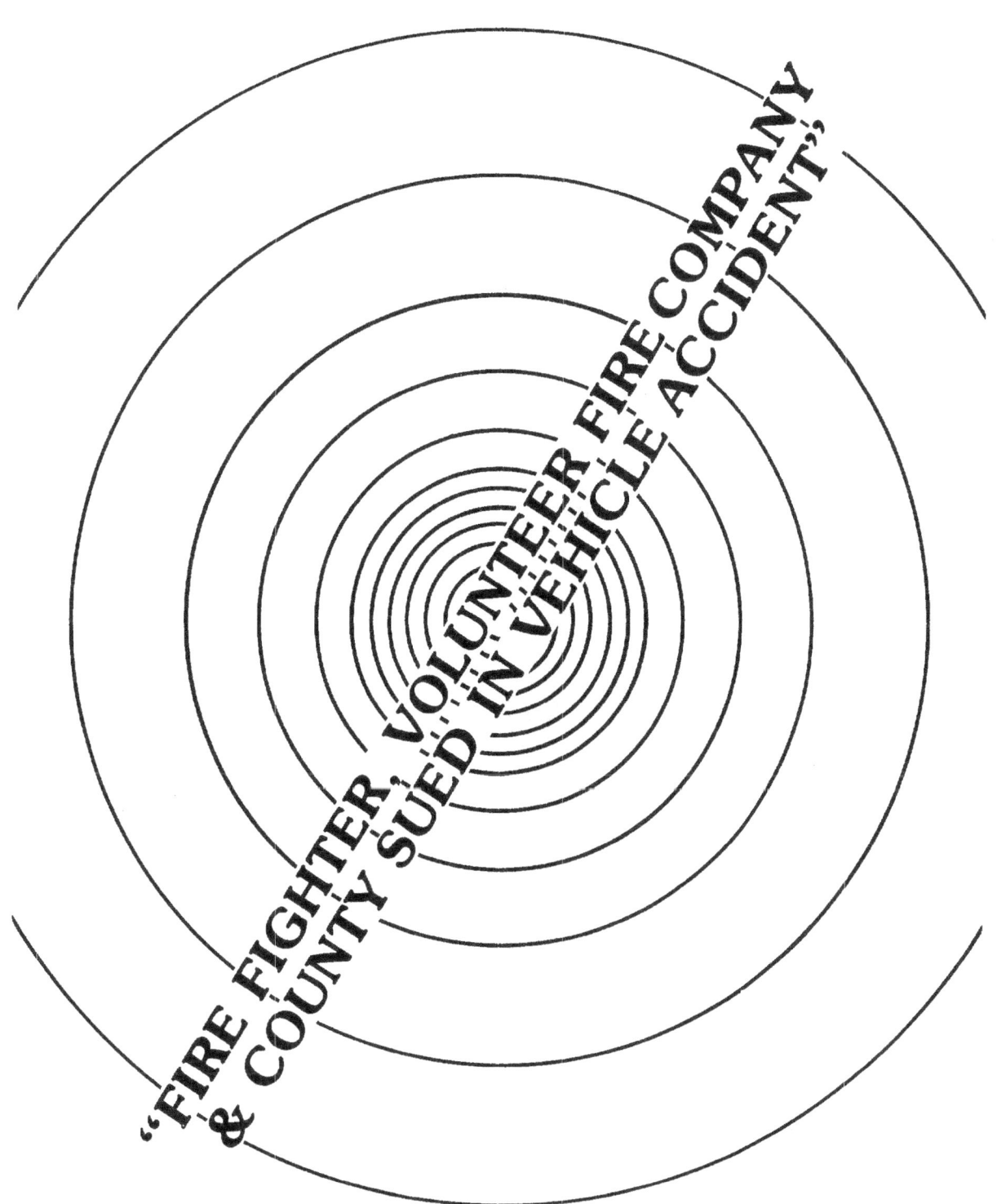

"FIRE FIGHTER, VOLUNTEER FIRE COMPANY & COUNTY SUED IN VEHICLE ACCIDENT;"

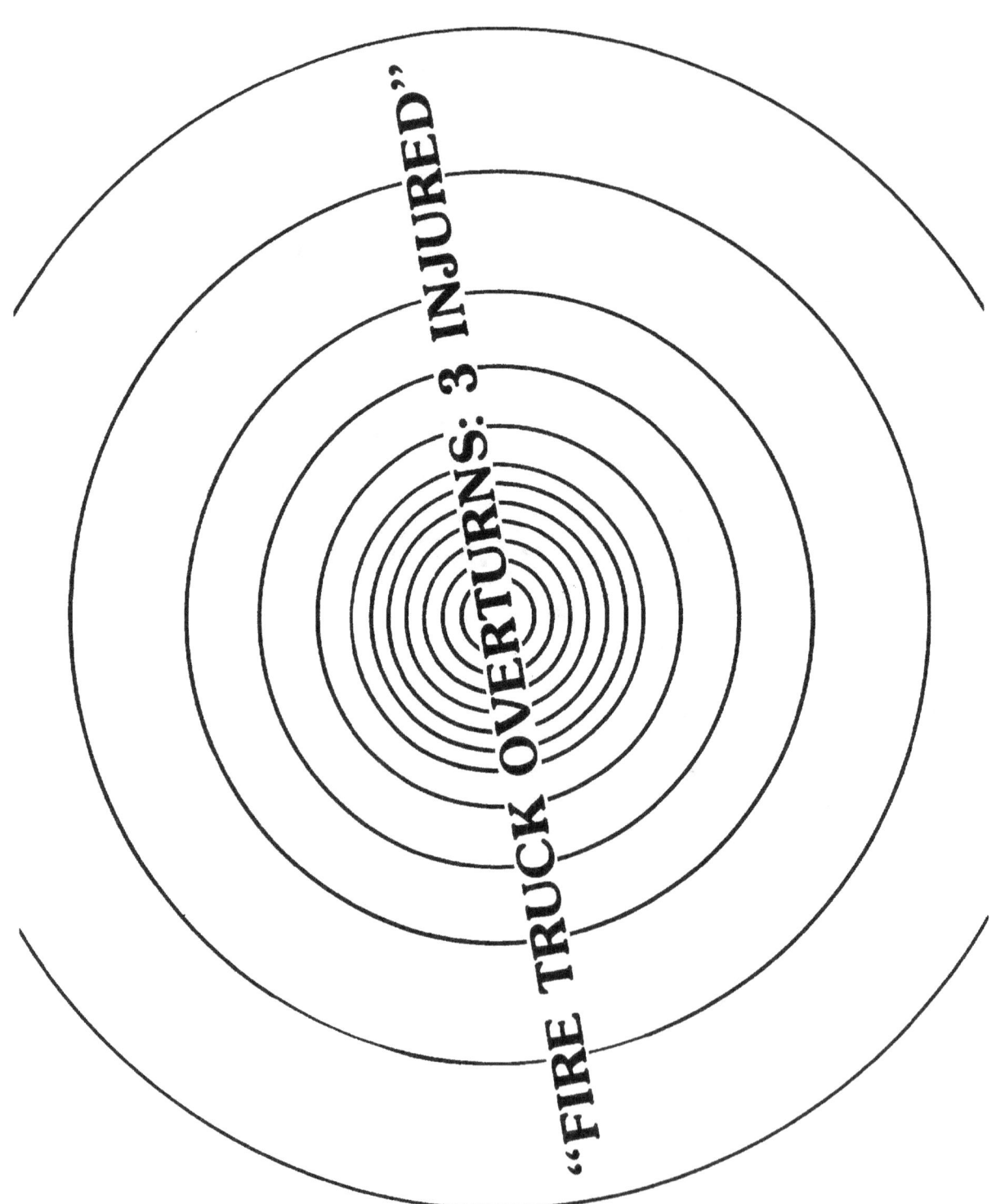

"FIRE TRUCK OVERTURNS: 3 INJURED"

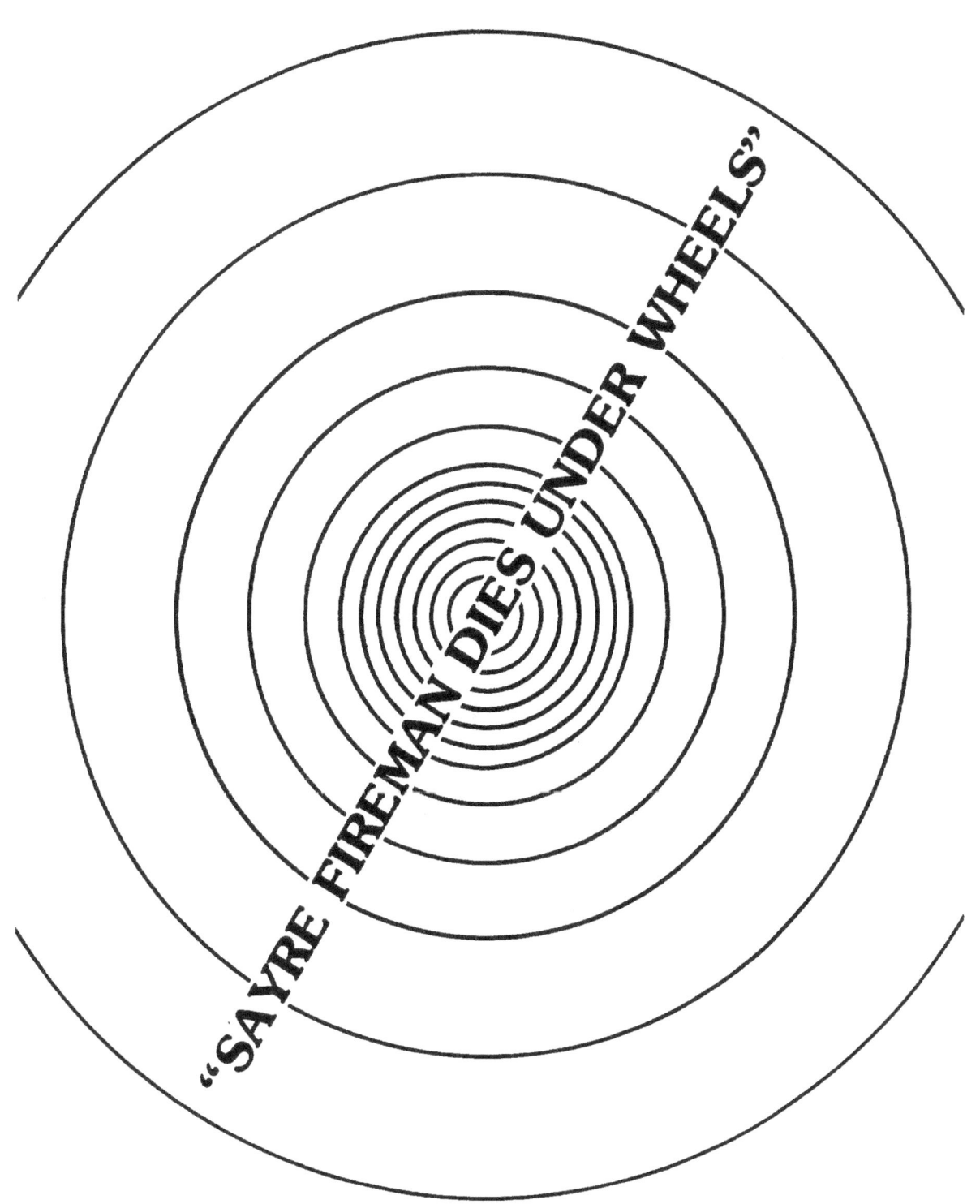

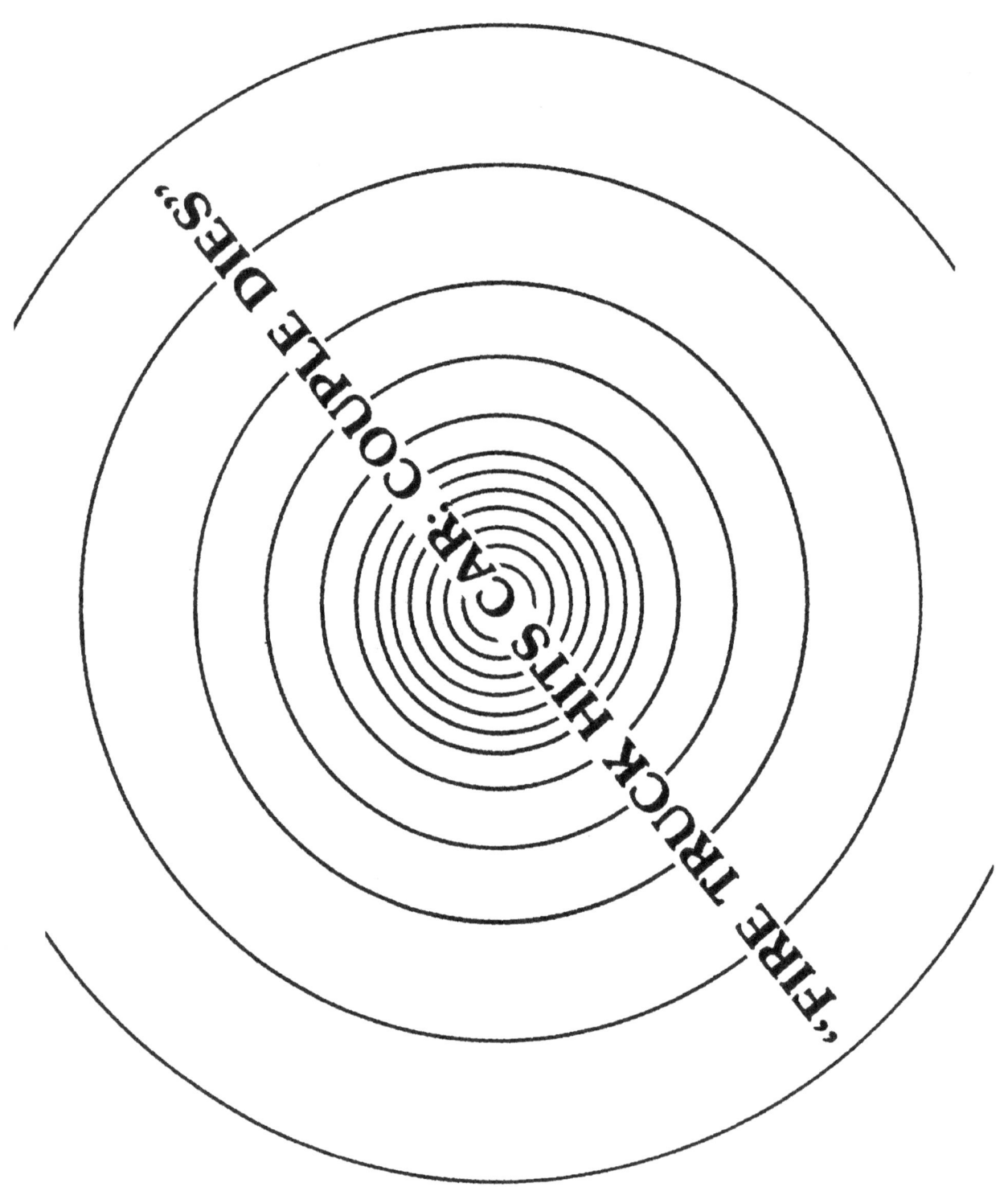

"FIRE TRUCK HITS CAR, COUPLE DIES."

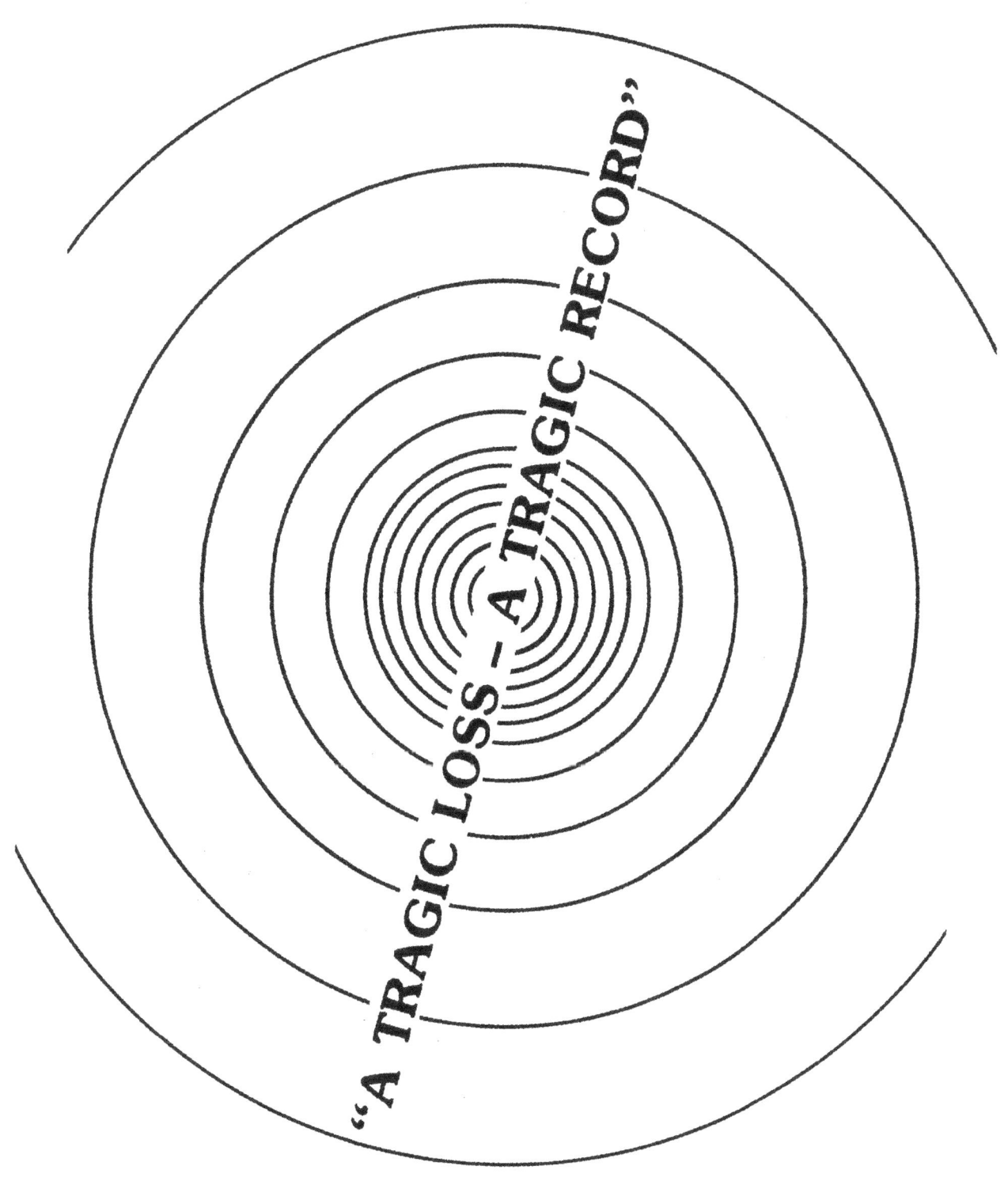

"A TRAGIC LOSS – A TRAGIC RECORD"

V.F.I.S. STATISTICAL INFORMATION

	PUMPERS/ TANKERS	AMBULANCE RESCUE SQUADS
INTERSECTION	16%	23%
INADEQUATE CLEARANCE - Fixed Object	45%	22%
INADEQUATE CLEARANCE - Moving Object	10%	18%

TOTAL NUMBER OF REPORTED ACCIDENTS FOR FIRST AND SECOND YEARS

⇨ FIRST YEAR

202	Ambulance Accidents
27.9	Accidents/100 Motor Vehicles
28.4	Accidents/Million Vehicle Miles

⇨ SECOND YEAR

226	Ambulance Accidents
29.7	Accidents/100 Motor Vehicles
31.1	Accidents/Million Vehicle Miles

VEHICLE INVOLVEMENT FOR FIRST AND SECOND YEARS

	FIRST YEAR	SECOND YEAR
	201 Accidents	210 Accidents
Single Vehicle	106 (52.7%)	112 (53.3%)
Multiple Vehicle	92 (45.8%)	90 (42.9%)
Pedestrian	3 (1.5%)	8 (3.8%)

AMBULANCE USE DURING TIME OF ACCIDENT FOR FIRST AND SECOND YEARS

Type of Use	FIRST YEAR 193 Accidents	SECOND YEAR 219 Accidents
Responding to a Call	55 (28.5%)	60 (27.4%)
Transferring Patient	59 (30.6%)	65 (29.7%)
Returning from a Call	48 (24.9%)	56 (25.6%)
Other Use	31 (16.1%)	38 (17.4%)

LIGHT CONDITIONS AT TIME OF ACCIDENTS
FOR FIRST AND SECOND YEARS

Light Conditions	FIRST YEAR 189 Accidents	SECONDYEAR 212 Accidents
Dawn/Dusk	19 (10.1%)	23 (10.8%)
Daylight	115 (60.8%)	108 (50.9%)
Dark	34 (18.0%)	58 (27.4%)
Unknown	21 (11.1%)	23 (10.8%)

ROAD SURFACE CONDITIONS AT TIME OF ACCIDENTS FOR FIRST AND SECOND YEARS

Surface Conditions	FIRST YEAR	SECOND YEAR
	196 Accidents	213 Accidents
Dry	97 (49.5%)	130 (61.0%)
Wet	26 (13.3%)	22 (10.3%)
Snow/Ice	27 (13.8%)	28 (13.1%)
Muddy	0 (0.0%)	1 (0.5%)
Unknown	46 (23.5%)	32 (15.0%)

TYPE OF COLLISION FOR
FIRST AND SECOND YEARS

Collision With	FIRST YEAR	SECOND YEAR
	190 Accidents	222 Accidents
Another Vehicle in Transport	62 (32.6%)	61 (27.5%)
Parked Vehicle	32 (16.8%)	46 (20.7%)
Fixed Object	71 (37.4%)	76 (34.2%)
Pedestrian	3 (1.6%)	8 (3.6%)
Animal	4 (2.1%)	5 (2.2%)
Non-Collision	0 (0.0%)	8 (3.6%)
Other	18 (9.5%)	18 (8.1%)

ACTION OF THE AMBULANCE AND ITS OPERATOR AT THE TIME OF THE ACCIDENT FOR FIRST AND SECOND YEARS

Operator's Actions	FIRST YEAR 157 Accidents	SECOND YEAR 174 Accidents
Backing Vehicle	60 (38.2%)	61 (35.1%)
Turning Vehicle Around	16 (10.2%)	21 (12.1%)
Approaching Intersection	10 (6.4%)	9 (5.2%)
Proceeding Straight/ Intersection	28 (17.8%)	30 (17.2%)
Turning/Intersection	14 (9.0%)	11 (6.3%)
Entering/Exiting an Expressway	1 (0.6%)	1 (0.6%)
Passing	13 (8.3%)	24 (13.8%)
In a Curve	10 (6.4%)	11 (6.3%)
Parking	3 (1.9%)	3 (1.7%)
Other	2 (1.3%)	3 (1.7%)

CONTRIBUTING FACTOR ASSOCIATED WITH THE AMBULANCE AND/OR OPERATOR FOR ACCIDENTS DURING FIRST AND SECOND YEARS

Contributing Factors	FIRST YEAR 157 Accidents	SECOND YEAR 183 Accidents
No Violation or Failure	107 (68.2%)	122 (66.7%)
Skidding/Loss of Control	18 (11.5%)	16 (8.7%)
Failing to Yield	17 (10.8%)	19 (10.4%)
Improper Passing	7 (4.5%)	8 (4.4%)
Following Too Closely	5 (3.2%)	6 (3.3%)
Too Fast for Conditions	1 (0.6%)	5 (2.7%)
Reckless Driving	1 (0.6%)	4 (2.2%)
Misjudging Driving	1 (0.6%)	1 (0.5%)
Vehicle Failure	0 (0.0%)	2 (1.1%)

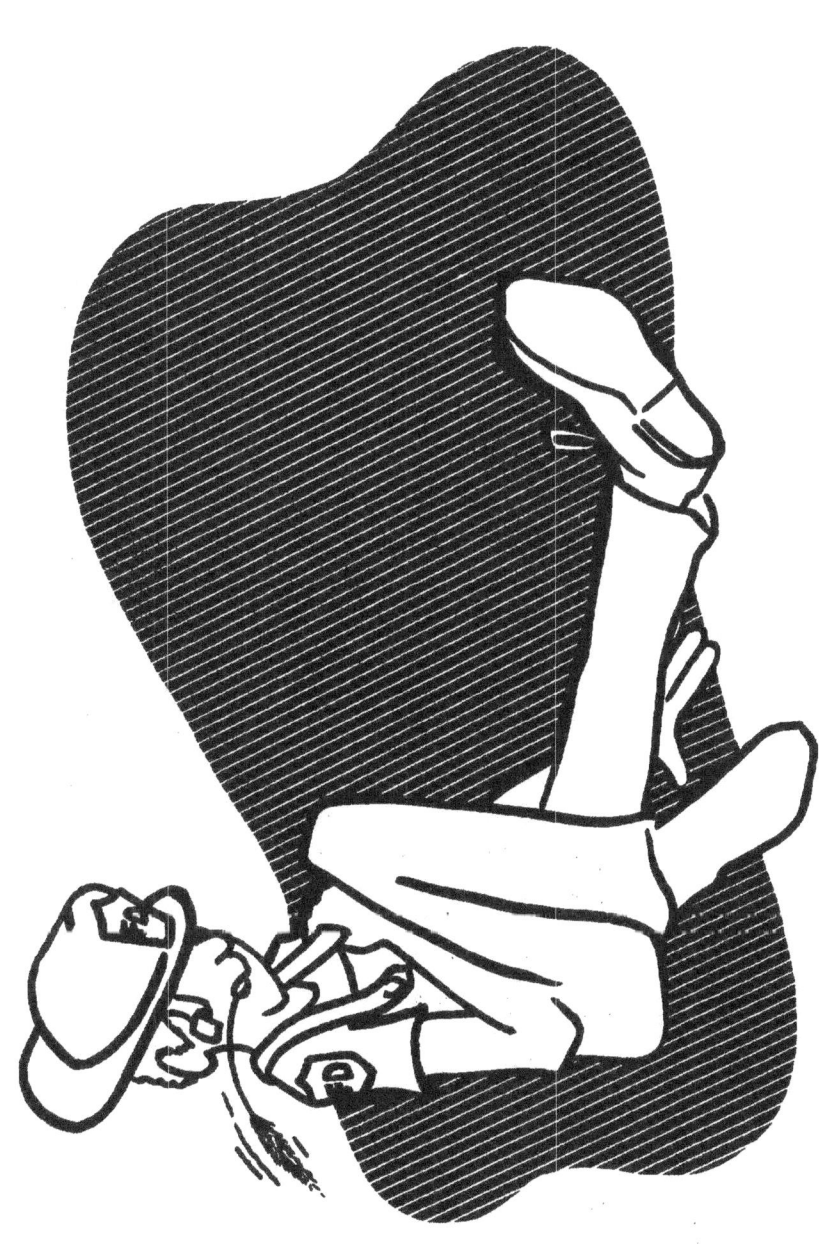

HUMAN ASPECTS

- Attitude
- Knowledge
- Mental Fitness
- Judgment
- Physical Fitness
- Age
- Habits
- Driving Characteristics

ACQUIRED ABILITY

- Driver's License
- State and Local Laws
- Defensive Driving Techniques
- Vehicle Characteristics
- Handling Vehicle

TRUE EMERGENCY . . . .

A situation in which there i's a high probability of death or serious injury to an individual, or significant property loss, and action by an Emergency Vehicle operator may reduce the seriousness of the situation.

DUE REGARD . . . .

'Enough' notice of approach, before a collision is inevitable.

Velocity

Reaction

Centripetal Force

Inertia

Centrifugal Force

Friction

Momentum

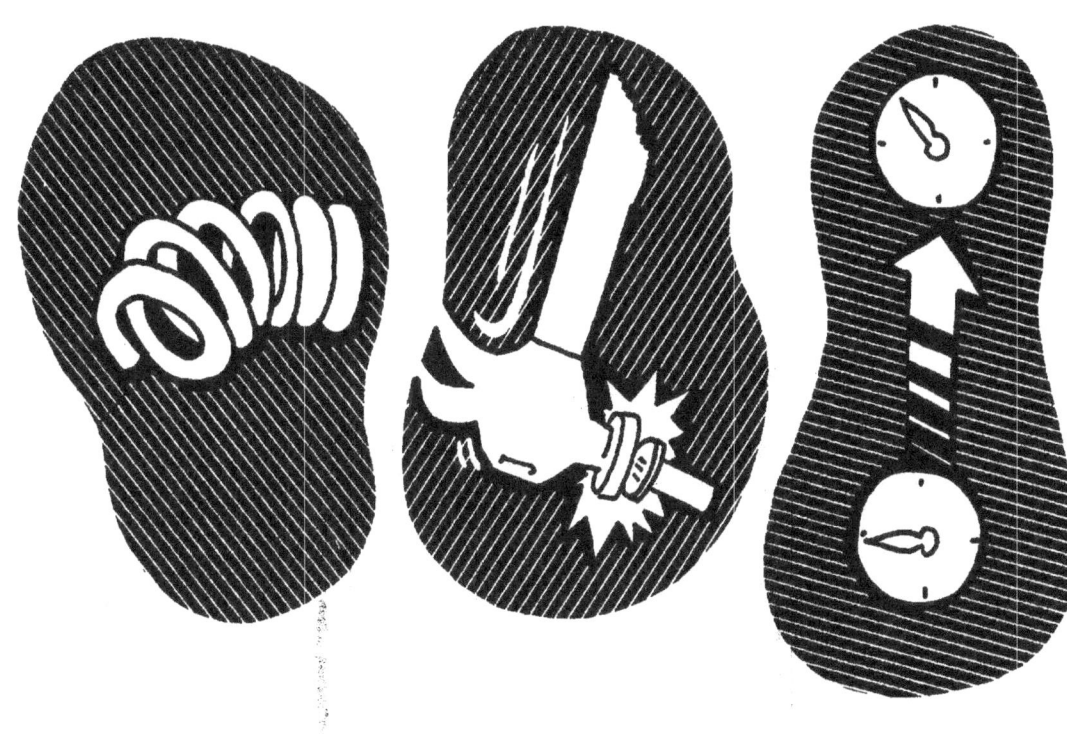

Potential Energy

Kinetic Energy

Velocity

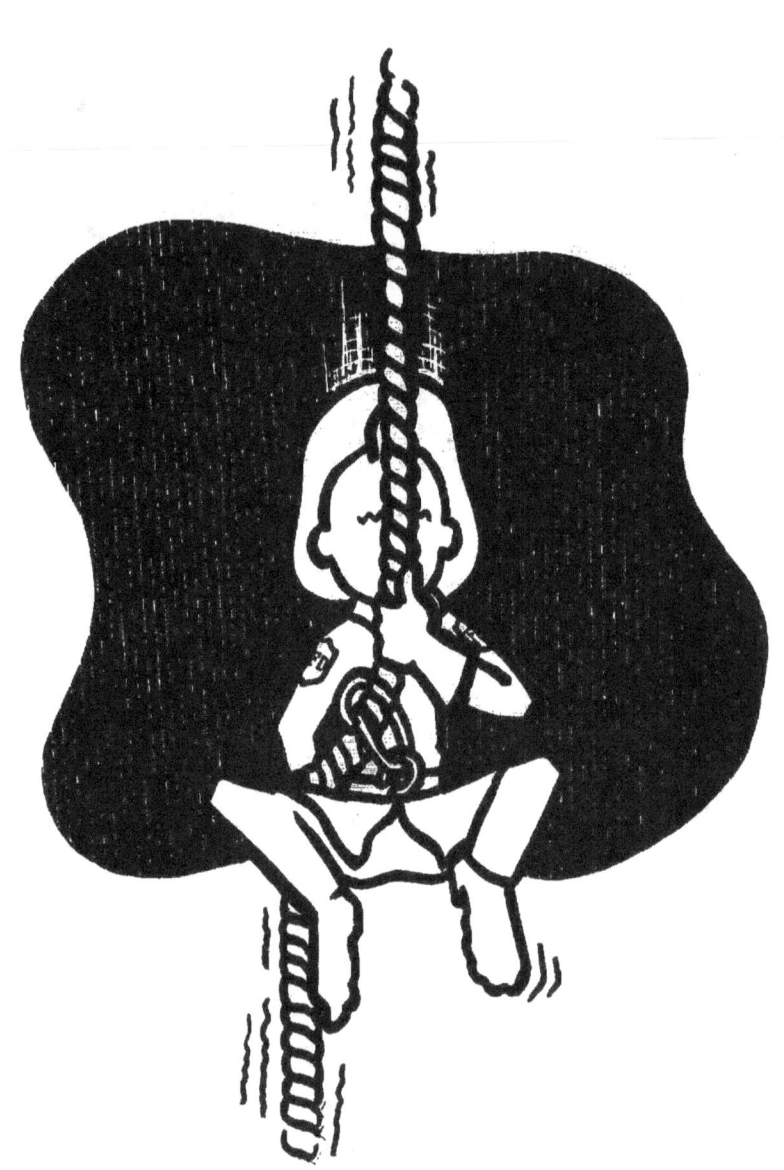

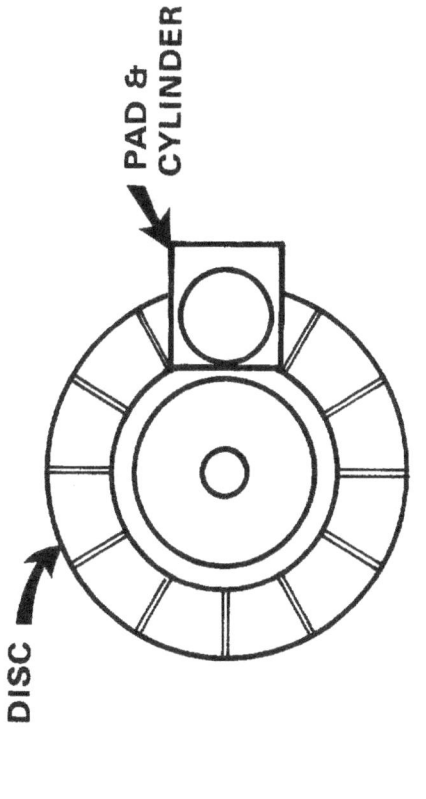

DISC

PAD & CYLINDER

DISC BRAKE

DRUM

BRAKE LINING

BRAKE SHOES

DRUM BRAKE

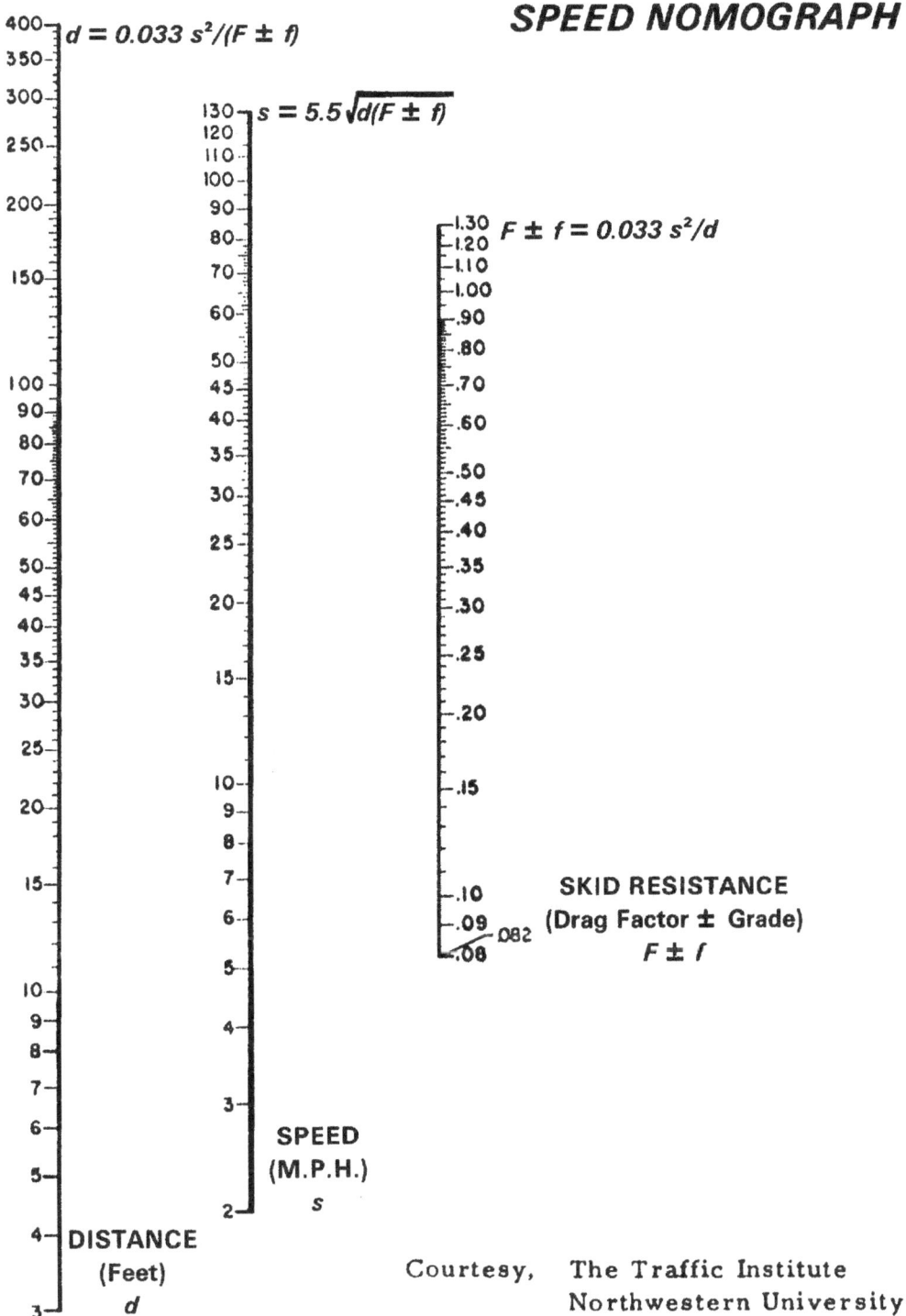

SKIDMARK
SPEED NOMOGRAPH

$d = 0.033\ s^2/(F \pm f)$

400
350
300
250
200

150

100
90
80
70
60

50
45
40

35
30

25

20

15

10
9
8
7

6

5

4

3

DISTANCE
(Feet)
d

$s = 5.5\sqrt{d(F \pm f)}$

130
120
110
100
90
80
70
60
50
45
40
35
30
25

20

15

10
9
8
7
6
5

4

3

2

SPEED
(M.P.H.)
s

$F \pm f = 0.033\ s^2/d$

1.30
1.20
1.10
1.00
.90
.80
.70
.60
.50
.45
.40
.35
.30
.25
.20
.15
.10
.09   .082
.08

SKID RESISTANCE
(Drag Factor ± Grade)
$F \pm f$

Courtesy,   The Traffic Institute
Northwestern University

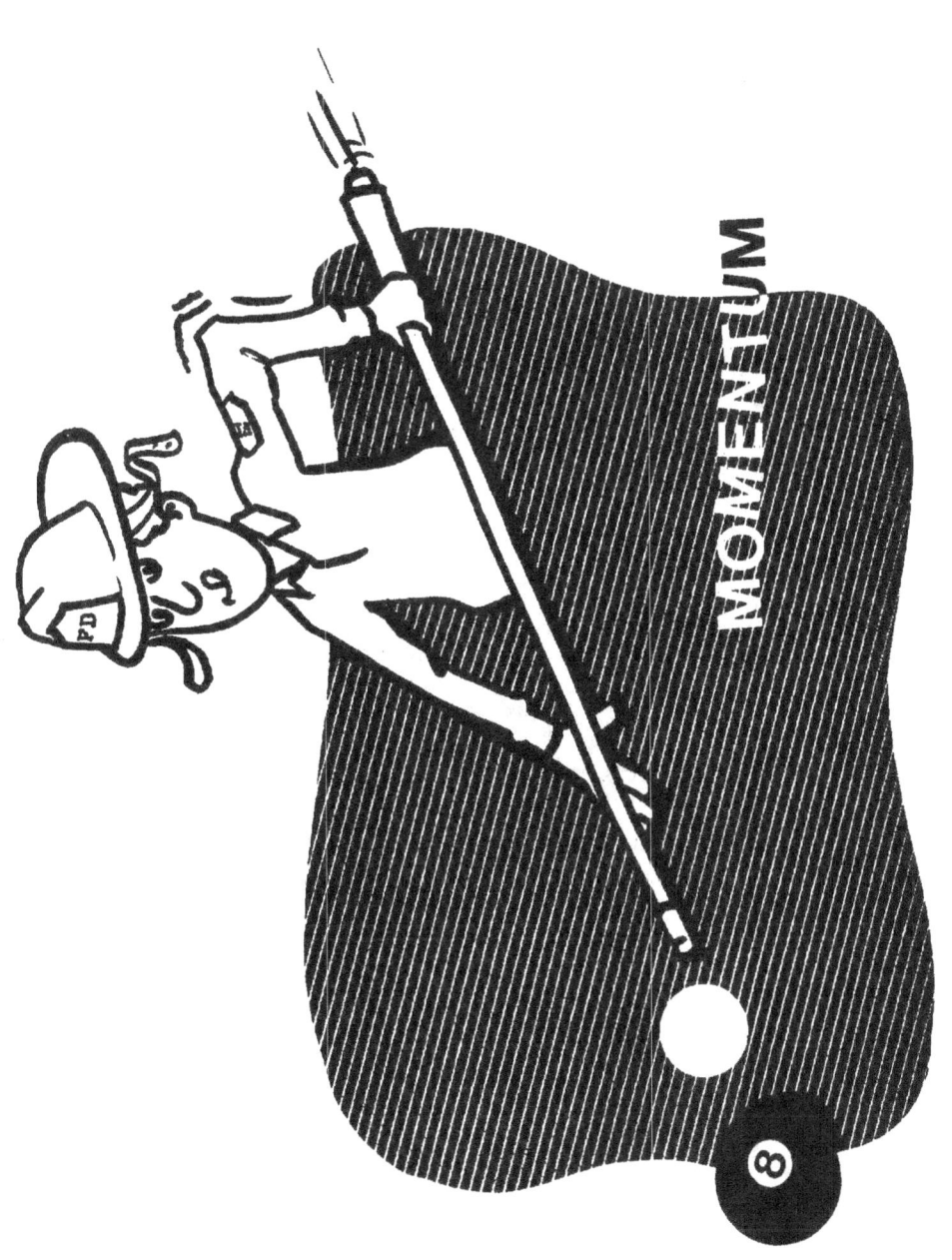

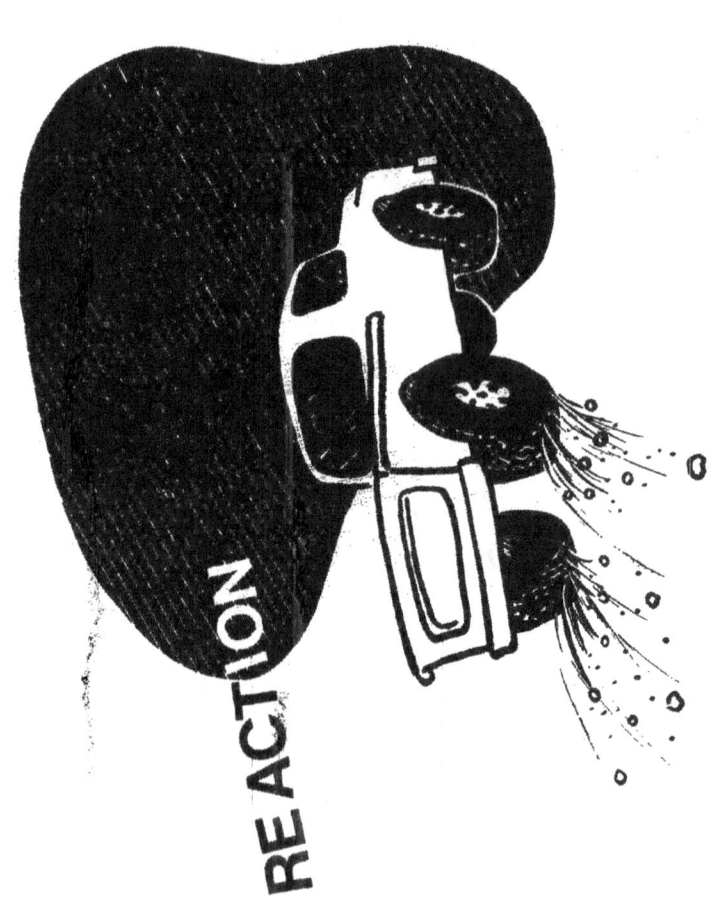

RE ACTION

AMBULANCE:

$$\frac{mv^2}{r}$$

Mass = 3,000 pounds = $\frac{W}{g}$ =

$$\frac{3,000 \text{ pounds}}{32.2 \text{ ft/sec}^2} = 93 \text{ slugs}$$

Velocity = 60 MPH to ft/sec by;
MPH (5280 ft per mile/
3600 sets per hour) =
88 ft/second.

Radius = 500 feet

C.F. for Ambulance =

$$\frac{(93 \text{ slugs})(88 \text{ ft/sec})^2}{500 \text{ foot radius}} =$$

1,440 pounds = 1,400 pounds

PUMPER:

$$\text{Mass} = 30{,}000 \text{ pounds} = \frac{W}{g} =$$

$$\frac{30{,}000 \text{ pounds}}{32.2 \text{ ft/sec2}} = 932 \text{ slugs}$$

Velocity = 60 MPH to ft/sec by;
       MPH (5280 ft per mile/
       3600 secs per hour) =
       88 ft/second.

Radius = 500 feet

CF. for Ambulance =

$$\frac{(932 \text{ slugs})(SS \text{ ft/sec})^2}{500 \text{ foot radius}} =$$

14,434 pounds = 14,000 pounds

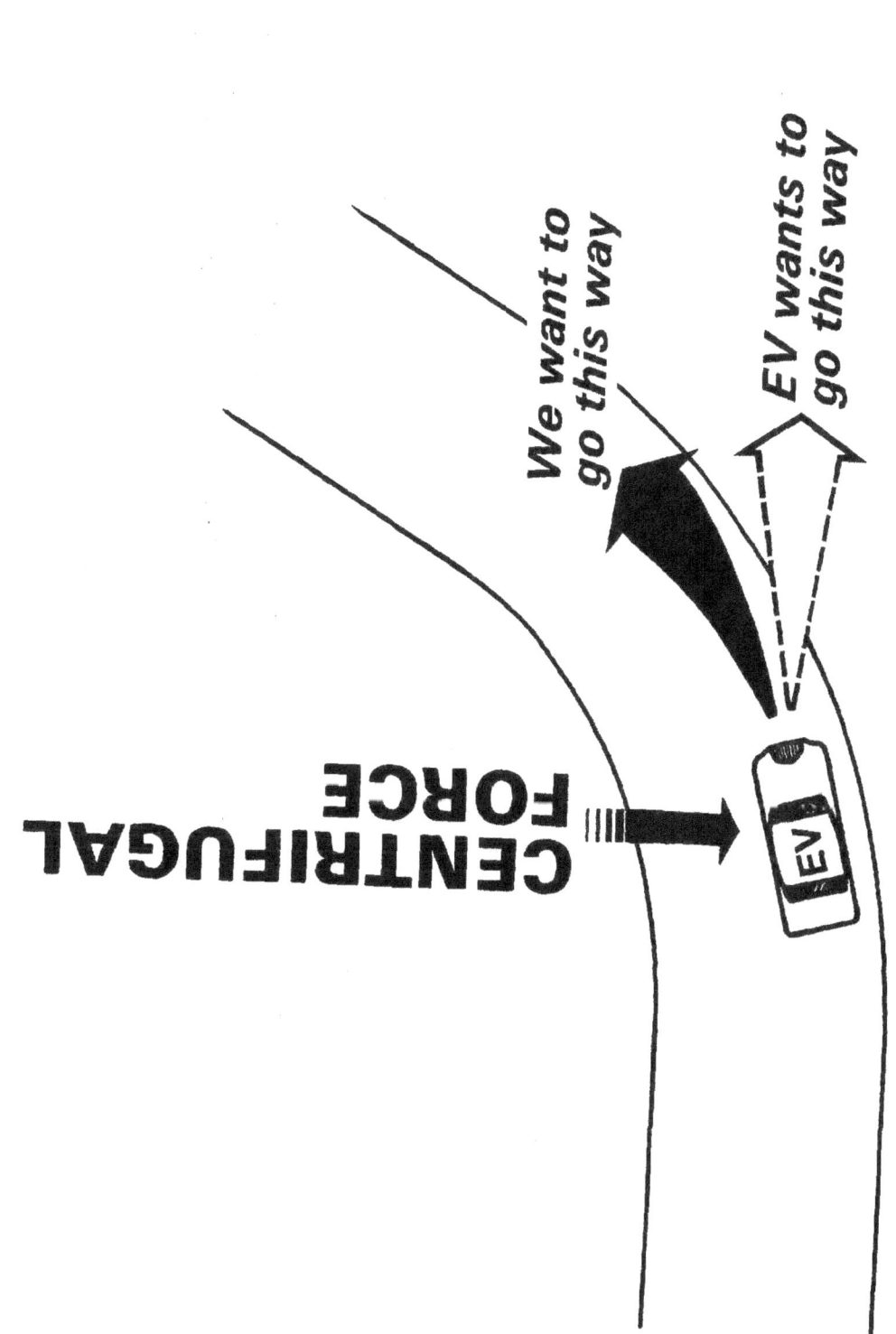

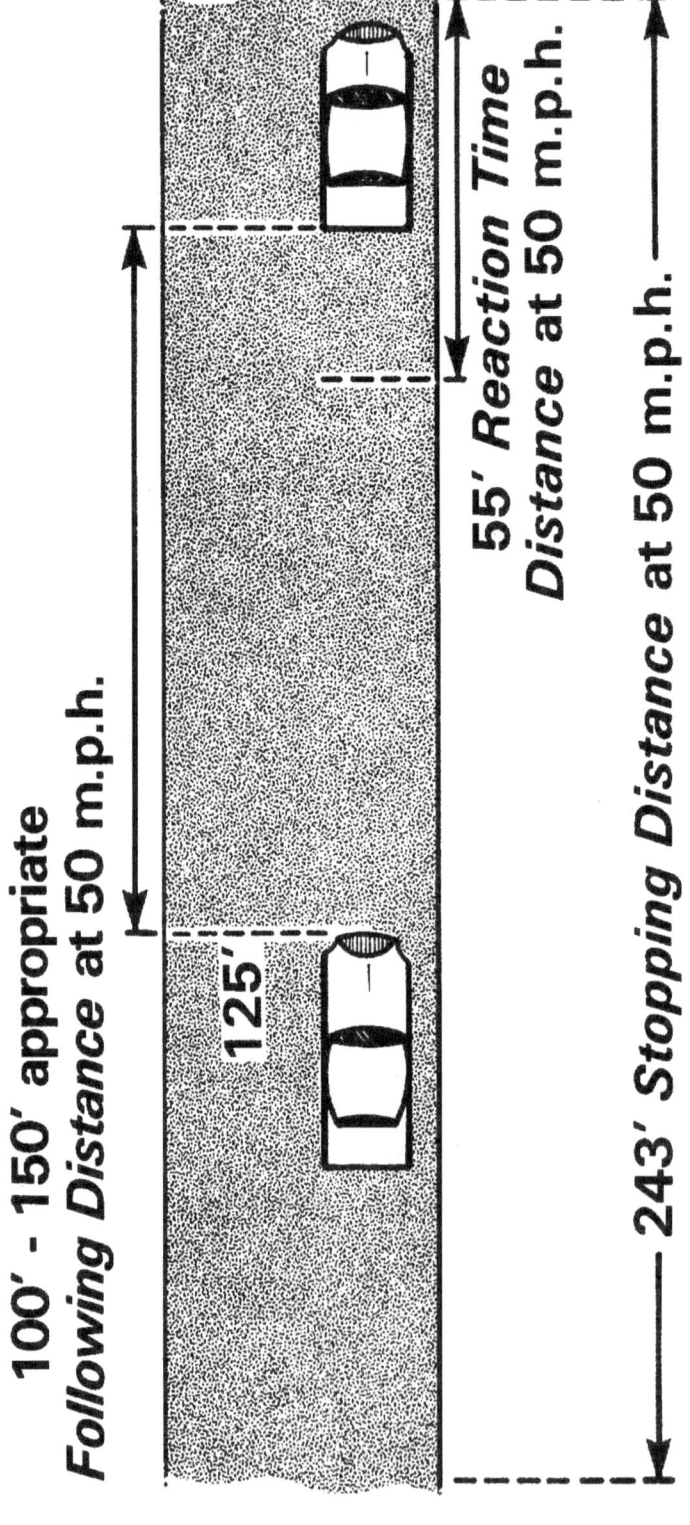

100' - 150' appropriate
Following Distance at 50 m.p.h.

125'

55' Reaction Time
Distance at 50 m.p.h.

243' Stopping Distance at 50 m.p.h.

LIGHT 2-AXLE TRUCKS

Speed Miles Per Hour	Feet Per Second	Driver # Reaction Distance	Vehicle # # Braking Distance	TOTAL * STOPPING DISTANCE
10	15	11	7	18
15	22	17	17	34
20	29	22	30	52
25	37	28	46	74
30	44	33 11 yds.	67 22 yds.	100 33 yds.
35	51	39	92	131
40	59	44 15 yds.	125 42 yds.	169 56 yds.
45	66	50	165	215
50	73	55 18 yds.	225 75 yds.	280 93 yds.
55	81	61	275	336
60	88	66 22 yds.	360 120 yds.	426 142 yds.

HEAVY 2-AXLE TRUCKS

Speed		Driver # Reaction Distance	Vehicle # # Braking Distance	TOTAL * STOPPING DISTANCE
Miles Per Hour	Feet Per Second			
10	15	11	10	21
15	22	17	22	39
20	29	22	40	62
25	37	28	64	92
30	44	33	92	125
35	51	39	125	164
40	59	44	165	209
45	66	50	210	260
50	73	55	255	310
55	81	61	310	371
60	88	66	370	436

3-AXLE TRUCKS AND COMBINATIONS

Speed		Driver # Reaction	Vehicle # # Braking	TOTAL * STOPPING
Miles	Feet	Distance	Distance	DISTANCE
Per Hour	Per Second			
10	15	11	13	24
15	22	17	29	46
20	29	22	50	72
25	37	28	80	108
30	44	33	115	148
35	51	39	160	199
40	59	44	205	249
45	66	50	260	310
50	73	55	320	375
55	81	61	390	451
60	88	66	465	531

3-AXLE TRUCKS AND COMBINATIONS

Speed		Driver # Reaction	Vehicle # # Braking	TOTAL * STOPPING
Miles Per Hour	Feet Per Second	Distance	Distance	DISTANCE
10	15	11	13	24
15	22	17	29	46
20	29	22	50	72
25	37	28	80	108
30	44	33	115	148
35	51	39	160	199
40	59	44	205	249
45	66	50	260	310
50	73	55	320	375
55	81	61	390	451
60	88	66	465	531

On impact, the car begins to crush and slow down. The person inside continues to move forward at the same speed the car was traveling.

Within 1/10 of a second, the car has come to a stop, but the person is still moving forward.

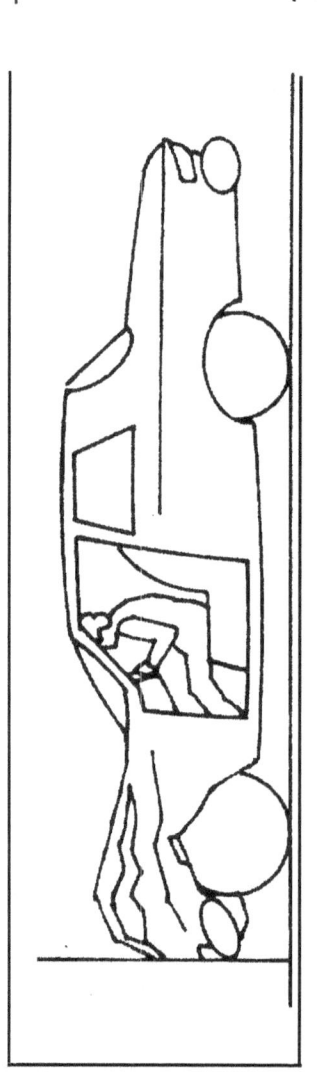

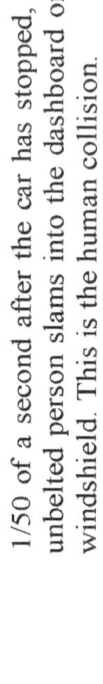

1/50 of a second after the car has stopped, the unbelted person slams into the dashboard or windshield. This is the human collision.

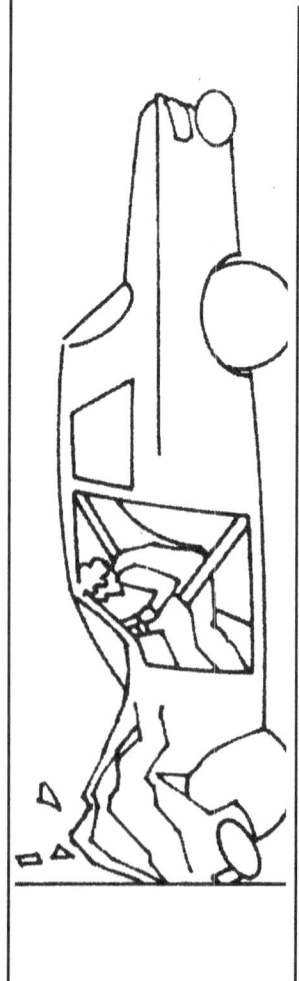

With effective safety belts, the person will stop before his or her head or chest hits the steering wheel, dash or windshield.

THE MYTHS

The National Highway Traffic Safety Administration discusses various myths regarding the reluctance of some to use seat belts. These should also be viewed concerning emergency vehicle drivers.

Myth	*Fact*
A. "I don't need seat belts because I'm a really good driver. I have excellent reactions. "	"No matter how good a driver you are, you can't control the other car. When another car comes at your, it may be the result of a mechanical failure and there's no way to protect yourself against someone else's poor judgment and bad driving."
B. "I don't want to be trapped in by a seat belt. It's better to be thrown free in an accident."	"Being thrown free is 25 times more dangerous . . . 25 times more lethal. If you're wearing your belt you're far more likely to be conscious after an accident . . . To free yourself and help your passengers."
C. "I just don't believe it will ever happen to me."	"Everyone of us can expect to be a crash once every 10 years for one out of 20 of us, it'll be a serious crash. For one out of every 60 born today, it will be fatal."
D. "Well, I only need to wear them when I have to go on long trips, or at high speeds."	"Eighty (SO) percent of deaths and serious injuries occur in cars traveling under 40 miles per hour and 75 percent of deaths or injuries occur less than 25 miles from your home."
E. "I can touch my head to the dashboard when I'm wearing my seat belt so there's no way it can help me in a car accident."	"Safety belts were designed to allow you to move freely in your car. They were also designed with a latching device that locks the safety belt in place if your car should come to a sudden halt. This latching device keeps you from hitting the inside of the car or being ejected. It's there when you need it."
F. "I don't need it. In case of an accident, I can brace myself with my hands."	"At 35 miles per hour, the force of impact on you or your passengers is brutal. There's no way your arms and legs can brace you against that kind of collision. The speed and force are just too great. The force of impact at just 10 MPH is equivalent to the force of catching a 200-pound bag of cement from a 1st-story window."
G. "Most people would be offended if I asked them to put on a seat belt in my car."	"Polls show that the overwhelming majority of passengers would even willingly put their own belts on if only you, the driver, would ask them."

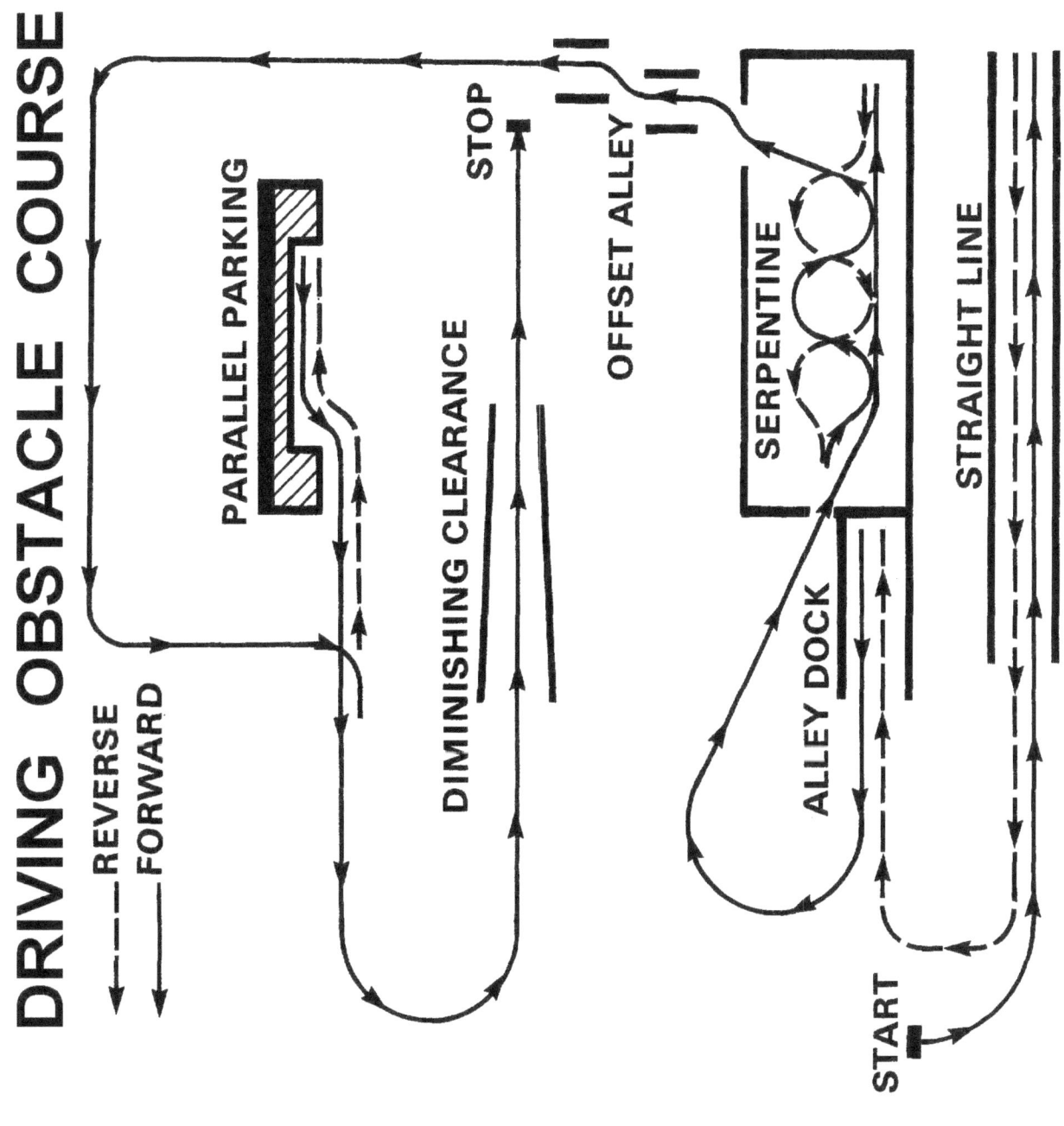

DRIVING OBSTACLE COURSE

REVERSE ----

FORWARD ——

PARALLEL PARKING

DIMINISHING CLEARANCE

STOP

OFFSET ALLEY

SERPENTINE

STRAIGHT LINE

ALLEY DOCK

START

INDEX

INDEX

Acceleration, 69, 70, 74
Aerial Truck Accident, 20, 21
Ambulance Standard Operating Procedure (SOP), Appendix B
Appendices:
    State Laws/Local Statutes, Appendix A
    Scarsdale Ambulance SOP, Appendix B
    VFIS Driving Course, Appendix C
    Instructors Key To Student Workbook, Appendix D
    Bibliography, Appendix E

Bibliography, Appendix E
Brakes, 74
    Braking Distance, 86-92
    Changing Direction, 74-75
    Disc, 73
    Drum, 73
    Fade, 73
    Friction at, 72
    Friction Nomograph, 75, 77
    Locking the Wheels, 74
    Velocity Control & Friction, 74-75

Centrifugal Force, 82-84
Centripetal Force, 85
Charts:
    Driver Training Test/Instruction-Scarsdale S.O.P., Appendix B
    Friction Nomograph, 77
    Pavement Drag Factors for Rubber Tires, 76
    Total Stopping Distance
        Light 2-Axle Trucks, 91
        Heavy 2-Axle Trucks, 91
        3-Axle Trucks & Combinations, 92

Distances, 85-92
    Considerations During Emergency Mode, 89-90
    Charts:
        Light 2-Axle Trucks, 91
        Heavy 2-Axle Trucks, 91
        3-Axle Trucks & Combinations, 92
    Following, 85-92
        Estimating, 86-89
        What is Safe? 86

Increasing, 88-89
    By 50%, 88
    Double the Distance, 89
    Triple the Distance, 89
Judging, 88
    Estimating Car Lengths, 88
    "Two-Second Rule," 88
Stopping, 86, 90
    Braking Distance, 86
    For Various Vehicle Types, 87, 89
    Reaction Distance, 86
Drivers:
Concerns, 1-3
Impact on Families, 3
Injury to Others, 2
Loss of Equipment, 2
Personal Injury, 2
Personnel Selection, 2, 41-53
Suit Filed Against, 16-17

Energy:
Kinetic, 68
Potential, 68
Engine Accidents:
On Maiden Run, 9-10
Overturn: 3 Injured, 18-19
Three Hurt, 11-13
"Tragic Loss-Tragic Record," 25-26

Fatalities:
Elderly Couple with Tanker, 22-24
Fire Fighter and Aerial Truck, 20-21
Five Fire Fighters, 14-15
Suit Filed Against Driver, 16-17
Forms:
Scarsdale SOP: Driver Training Progress Report, Appendix B
YFIS Personnel File Sheet, 52
VFIS Preventive Maintenance Form
Formulae:
Centrifugal Force, 84
Coefficient of Friction, 75
Law of Conservation of Momentum, 81
Law of Momentum, 79
Law of Momentum for Different Masses (weights), 82
Minimum Following Distance:
    "Two-Second Rule," 88
    "Three-Second Rule," 88
Friction, 70-77
At the Brakes, 72
Changing Direction and, 74-75

Coefficient of, 75-76
Tire and Road, 91
Velocity Control and, 74-75

Incident Histories (News Articles), 9-26
    Elderly Couple's Death with Tanker, 22-24
    Engine Accident: 3 Hurt, 11-13
    Engine on Maiden Run, 9-10
    Engine Overturn: 3 Injured, 18-19
    Fire Fighter Fatality: Aerial Truck, 20-21
    Five Fire Fighters Killed, 14-15
    Suit Filed Against Driver, 16-17
    "Tragic Loss-Tragic Record," 25-26
identifying;
    Distances, 85-92
        Braking, 86-87
        Reaction, 86
        Stopping, 87, 90-92
    The Problem, 1-4
Indiana State University of Pennsylvania Study: Synopsis, 28
inertia:
    and Momentum, 80-81
    Newton's Law, 77-79

Kinetic Energy, 68

Laws: Vehicle Code, Appendix A
Legal Aspects, 56-63
    Case History, 63
    "Good Faith," 58
    Immunity, 57
    Negligence, 58
    of Emergency Vehicle Operations, 57-58
    State Laws/Local Statutes, Appendix A
    Suggested Readings, 60
    Suit Filed Against Driver, 16-17
    What is a True Emergency? 62
Loss Statistics:
    VFIS Loss Statistics. 26-27

Momentum:
    and Intertia, 80
    Conservation of, 81
    Newton's Law, 79
Motivation, 33-37

Newton's Laws, 77-85
    Conservation of Momentum, 81
    Inertia, 77
    Momentum, 79
Momentum and Inertia, 80
Reaction, 82

Personnel, 41-53
    Acquired Ability, 44-46
    Discussion Topics, 49-5 1
    File, 46-49
    Human Aspects, 43-44
    Sample File Sheet, 52
    Selection, 41-46
Physical Forces, 67-92
    Acceleration, 69, 70, 74
    Centrifugal Force, 82-84
    Centripetal Force, 85
    Friction, 70-77
        At the Brakes, 72
        Brake Fade, 73
        Coefficient of, 75-76
        Tire and Road, 91
    Kinetic Energy, 68
    Potential Energy, 68
    Velocity, 68-69
    Velocity Control and Friction, 74
        Accelerating, 74
        Braking, 74
        Changing Direction, 74-75
        Locking the Wheels, 74
Potential Energy, 68

Reaction, 82
Reaction Distance, 86-87

Seat Belts, 92.1-92.7
    Belted Occupants, 92.5
    Crash Dynamics:
        Car's Collision, 92.3
        Human Collision, 92.3
        Factors Contributing to Injury and Death, 92.4
    Myths vs. Facts, 92.5-92.6
    National Highway Traffic Safety Administration Findings, 92.1
    Standards:
        49 CFR 571.208, Occupant Crash Protection, 92.2
        OSHA Standard Section 1915.100 (Proposed), 92.2
    Unrestrained Occupants, 92.4
Student Workbook:
    Instructor Key, Appendix D

Vehicle:
    Characteristics, 45-46
        Brakes, 72
    Control, 87
    Maintenance & Records (PM), 95-99
    Standard Operating Procedure, 103-106
Velocity, 68-69, 79
    Accelerating, 70
    Braking, 74-75
    Changing Direction, 74-75
    Control and Friction, 74-75
    Locking the Wheels, 74
VFIS:
    Loss Statistics, 26-27
    Personnel File Sheet, 52
    Preventive Maintenance, 95-99

FA-110 9/91

Student Workbook

United States Fire Administration

Emergency
Vehicle
Driver
Training

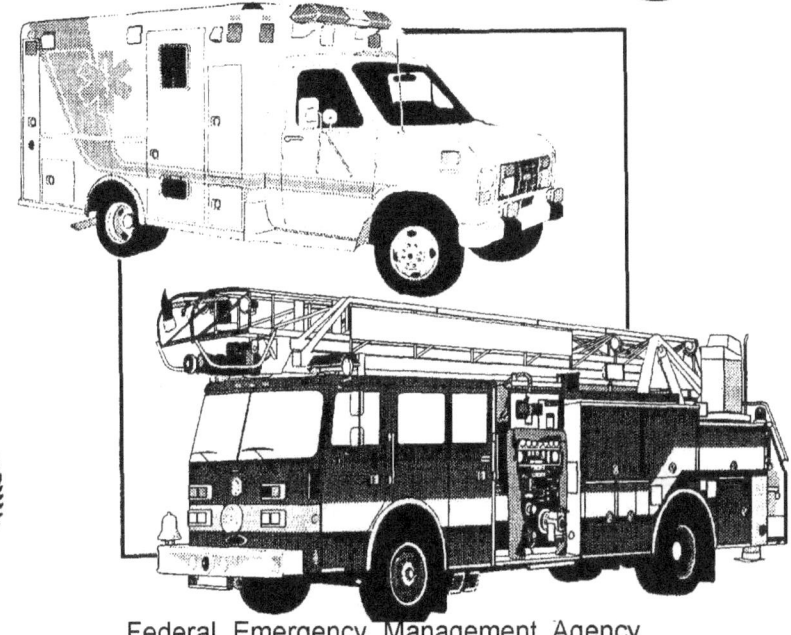

Federal Emergency Management Agency

Section I—Introduction

A. To what or whom must the driver of an emergency vehicle be responsible?

B. Of the six (6) prime factors which may cause vehicle accidents, list four (4)

   1. _____

   2. _____

   3. _____

   4. _____

C. Every emergency vehicle accident has the potential of causing losses that may vary in degree of severity from costing a few dollars to extreme disaster. List the four (4) areas in which this loss could be classified.

   1. _____

   2. _____

   3. _____

   4. _____

Section II—Identifying the Problem

This portion cites problems found from investigating particular vehicle accidents. It asks you to recall a few particulars from each case history.

A. What problems may a driver face if he is alone on the apparatus, responding to an emergency? _____

_____

B. What precautions must be exercised when approaching and entering an intersection during response?

   1. _____

     _____

   3. _____

C. Citizens may not react properly upon the approach of an Emergency Vehicle. List five (5) reasons for this:

   1. _____

   2. _____

   3. _____

   4. _____

   5. _____

D. Scenario Number One

_____

_____

_____

_____

   2. Peripheral injury or death

_____

_____

_____

_____

3. Equipment loss

_____
_____
_____
_____

4. Long term impact

_____
_____
_____
_____

E. Scenario Number Two

_____
_____
_____
_____

2. Peripheral injury or death

_____
_____
_____
_____

3. Equipment loss

_____
_____
_____
_____

4. Long term impact

_____
_____
_____
_____

F. Scenario Number Three

_____
_____
_____
_____

2. Peripheral injury or death

_____

_____

_____

_____

3. Equipment loss

_____

_____

_____

_____

4. Long term impact

_____

_____

_____

_____

G. Scenario Number Four

_____

_____

_____

_____

2. Peripheral injury or death

_____

_____

_____

_____

3. Equipment loss

_____

_____

_____

_____

4. Long term impact

_____

_____

_____

_____

H. Volunteer Firemen's Insurance Services Statistical Information

	PUMPERS/TANKERS	AMBULANCE RESCUE SQUADS
INTERSECTION		
INADEQUATE CLEARANCE —Fixed Object		
INADEQUATE CLEARANCE —Moving Object		

I. Notes From The Study By Indiana University of Pennsylvania

Section III—Motivation

A. Can vehicle drivers be motivated solely from an outside source or influence? Or must the driver possess an inner desire or drive to conduct himself safely? Discuss.

B. What two (2) needs are required to ensure an Emergency Vehicle driver can/does operate safely?

1. _____

2. _____

C. Discuss some areas within your agency's operation where you feel it may be time to change physical and psychological needs. (Be objective, not just critical!)

Section IV—Personnel Selection

A. Upon whom does the ultimate driver selection responsibility rest? Why?

1. _____

2. _____

B. Human Aspects: These ingredients are those provided by the prospective driver. They are the product of the constituent parts of this individual and how they will influence his abilities in either a positive or negative sense.

1. ATTITUDE: _____

_____

2. KNOWLEDGE: _____

_____

3. MENTAL FITNESS: _____

_____

4. JUDGEMENT: _____

_____

5. PHYSICAL FITNESS: _____

_____

6. AGE: _____

_____

7. HABITS: _____

_____

8. DRIVING CHARACTERISTICS: _____

_____

C. ACQUIRED ABILITY: These are abilities learned and demonstrated to agencies which sanction your performance.

1. DRIVER'S LICENSE: _____

_____

2. STATE AND LOCAL LAWS: _____

_____

3. DEFENSIVE DRIVING TECHNIQUES: _____

_____

Section V—Legal Aspects

A. What does the term, "negligence" mean? How does negligence figure in emergency vehicle operation?

_____

_____

_____

B. Emergency Vehicle operators are subject to all traffic regulations unless a specific exemption is made.

1. Indicate an exemption concerning the speed limit. _____

_____

_____

2. What mode must be present for the exemption to be legal? _____

_____

3. Even with an exemption to the driving laws, may the driver be found criminally or civilly liable? _____

C. If your actions are analyzed by legal authorities, they will be viewed from the aspects of; 1) was it a true emergency, and 2) was due regard exercised.

1. Discuss what is required to be defined as a true emergency. _____

_____

_____

2. Indicate what constitutes "due regard" in the operation of an Emergency Vehicle. _____

_____

_____

D. LEGAL ASPECTS OF EMERGENCY VEHICLE OPERATIONS: What are three types of regulations to follow?

_____

_____

3. _____

Section VI—Physical Forces

A. Physical forces are always present, acting on the vehicle. What are some driver actions which may result in lost vehicular control?

1. _____

2. _____

3. _____

4. _____

5. _____

B. While driving, the Emergency Vehicle operator can control velocity and direction only. Velocity control is the control over:

1. _____

2. _____

3. _____

C. Velocity is described as, "a rate of change of position, in relation to time." Give two examples of velocity and express the units of measurement.

1. _____

2. _____

D. Friction is the resistance to motion of two moving objects or surfaces that touch. Indicate three (3) places during Emergency Vehicle operation where friction exists.

1. _____

2. _____

3. _____

E. When is the friction between tires and the road least effective?

_____

_____

F. Brake fade is due to the generation of excessive heat. What braking technique can be used to reduce the possibility of this dangerous occurrence? _____

_____

G. FRICTION AT THE BRAKES: This friction is caused by the brake shoes pressing on the drums (or brake pads clamping the brake disc) to create a frictional contact to slow the wheel's turning. Notes on braking.

_____

_____

_____

_____

_____

_____

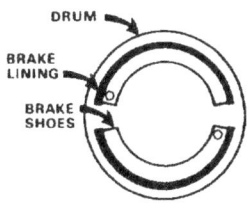

DRUM BRAKE

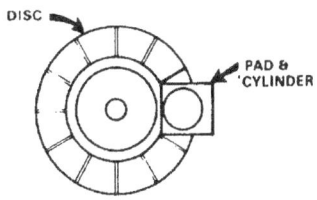

DISC BRAKE

POSSIBLE RANGES OF PAVEMENT DRAG FACTORS
FOR RUBBER TIRES

Description of Road Surface	DRY				WET			
	Less Than 30 m.p.h.		More Than 30 m.p.h.		Less Than 30 m.p.h.		More Than 30 m.p.h.	
	From	To	From	To	From	To	From	To
Cement								
New. Sharp	.80	1.00	.70	.85	.50	.80	.40	.75
Travelled	.60	.80	.60	.75	.45	.70	.45	.65
Traffic Polished	.55	.75	.50	.65	.45	.65	.45	.60
Asphalt								
New. Sharp	.80	1.00	.65	.70	.50	.80	.45	.75
Travelled	.60	.80	.55	.70	.45	.70	.40	.65
Traffic Polished	.55	.75	.45	.65	.45	.65	.40	.60
Excess Tar	.50	.60	.35	.60	.30	.60	.25	.55
Brick								
New. Sharp	.75	.95	.60	.85	.50	.75	.45	.70
Traffic Polished	.60	.80	.55	.75	.40	.70	.40	.60
Slone Block								
New. Sharp	.75	1.00	.70	.90	.65	.90	.60	.85
Traffic Polished	.50	.70	.45	.65	.30	.50	.25	.50
Gravel								
Packed. Oiled	.55	.85	.50	.80	.40	30	.40	.60
Loose	.40	.70	.40	.70	.45	.75	.45	.75
Cinders								
Packed	.50	.70	.50	.70	.65	.75	.65	.75
Rock								
Crushed	.55	.75	.55	.75	.55	.75	.55	.75
Ice								
Smooth	.10	.25	.07	.20	.05	.10	.05	.10
Snow								
Packed	.30	.55	.35	.35	.30	.60	.30	.60
Loose	.10	.25	.10	.20	.30	.60	.30	.60
Metal Grid								
Open	.70	.90	.55	.75	.25	.45	.20	.35

The drag factor or coefficient of a pavement, of a given description may vary considerably because quite a variety of road surfaces may be described in the same way and because of some variations due to weight of vehicle, air pressure in tire, tread design, air temperature, speed and some other factors.

These figures represent experiments made by many different people in all parts of the U.S. They are for straight skids on clean surfaces. Speeds referred to are at the beginning of the skid. This table is reproduced from the Accident Investigator's Manual for Police Published by the Traffic Institute Courtesy. The Traffic Institute
  Northwestern University

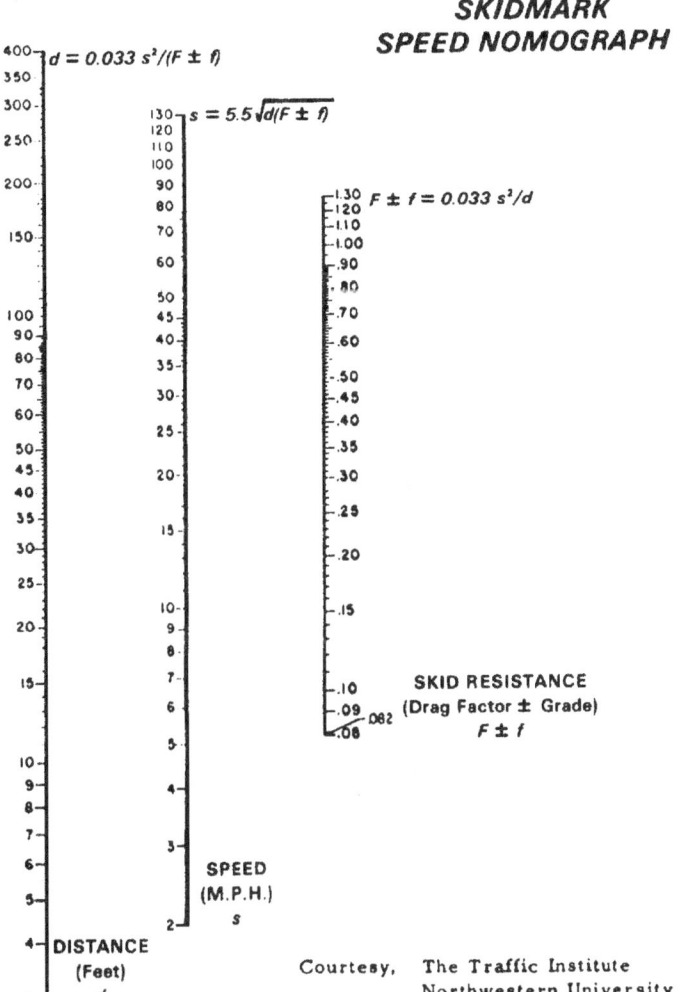

**SKIDMARK
SPEED NOMOGRAPH**

$d = 0.033\ s^2/(F \pm f)$

$s = 5.5\ \sqrt{d(F \pm f)}$

$F \pm f = 0.033\ s^2/d$

SKID RESISTANCE
(Drag Factor ± Grade)
$F \pm f$

SPEED
(M.P.H.)
s

DISTANCE
(Feet)
d

Courtesy,   The Traffic Institute
  Northwestern University

H. Newton's Law of Inertia states that "an object continues in its state of rest, or of uniform motion in a straight line, unless it is acted upon by a net external force." Applying this to your response area, what locations would be hazardous to you if you could not reduce inertia on your emergency vehicle?

1. _____

2. _____

3. _____

4. _____

I. The Law of Momentum states that, "when an unbalanced force acts on an object, the object will be accelerated. The acceleration will vary directly with the applied force and will be in the same direction as the applied force. It will vary inversely with the mass of the object." What potential damage may result at an intersection, if the vehicle's momentum is not reduced? _____

_____

_____

_____

J. "For every action there is an equal and opposite reaction," states the Law of Reaction. How does this apply to striking stationary objects? _____

_____

_____

_____

K. Compare the effects of Centrifugal force on a Pumper entering a 500 foot radius curve at 60 MPH and the effects on an ambulance entering the same curve at the same speed.

SOLUTION:

Centrifugal Force =

Where: m = mass, in pounds, to slugs
     r = radius, in feet
     v = velocity, in feet per second

CENTRIFUGAL FORCE

We want to go this way

EV wants to go this way

EV

AMBULANCE:

$$\text{Mass} = 3{,}000 \text{ pounds} = \frac{W}{g} = \frac{3{,}000 \text{ pounds}}{} = 93 \text{ slugs}$$

Velocity = 60 MPH to ft/sec by;
    MPH (5280 ft per mile/3600 secs per hour) =

Radius = 500 feet

$$\text{C.F. for Ambulance} = \frac{}{500 \text{ foot radius}} = 1{,}440 \text{ pounds} = 1{,}400 \text{ pounds}$$

8

PUMPER:

$$\text{Mass} = 30{,}000 \text{ pounds} = \frac{W}{g} = \frac{30{,}000 \text{ pounds}}{} = 932 \text{ slugs}$$

Velocity = 60 MPH to ft/sec by;
        MPH (5280 ft per mile/3600 secs per hour) =

Radius = 500 feet

$$\text{C.F. for Pumper} = \frac{}{500 \text{ foot radius}} = 14{,}434 \text{ pounds} = 14{,}000 \text{ pounds}$$

L. From the above problem, it can also be seen that a statement concerning Centrifugal Force can be made with regard to increasing a vehicle's weight, if other factors such as speed and radius of curve remain the same. Might this statement be: "If the vehicle's weight is increased ten times, Centrifugal Force increases ten times?" _____

M. If the speed for either the Pumper or Ambulance, above, is increased, will the Centrifugal Force realized increase or decrease? Why? _____

N. If the 500 foot curve's radius is decreased (the turn becomes sharper) will the Centrifugal Force be increased v or decreased? Why? _____

Supplemental Problems
Newton's Second Law

THE LAW OF MOMENTUM

"When an unbalanced force acts on an object, the object will be accelerated. The acceleration will vary directly with the applied force and will be in the same direction as the applied force. It will vary inversely with the mass of the object."

Formula to express this: $F = ma$

Where: F = Force, in pounds
        m = Mass, in slugs
        a = Acceleration, in feet/second^2

Mass is determined by: $m = \dfrac{W}{g}$

Where: m = mass in slugs
        W = Weight, in pounds
        g = Gravitational constant of 32.2 ft/sec^2

Acceleration is determined by converting MPH (miles per hour) into feet per second^2 by:

$$a = (MPH) \frac{5280 \text{ ft/mi.}}{3600 \text{ sec/hr}} =$$

EXAMPLE:

"What force is realized by a 30,000 pound engine traveling at 50 MPH?"

$$m = \frac{W}{g} = \frac{30{,}000}{32.2} = 932 \text{ slugs}$$

$$a = (50 \text{ MPH}) \frac{5280 \text{ ft/mile}}{3600 \text{ sec/hr}} =$$

F = ma
F =
F = 68,036 lbs

This formula of F = ma is simplified to determine momentum. It would now be expressed in the following formula:

$$M = mv$$

Where: M = Momentum in lb/sec
      m = mass, in slugs
      v = velocity, in ft/sec

Mass is obtained for this formula by: $\quad m = \dfrac{W}{g}$

Velocity is obtained by: $\quad v = (MPH) \dfrac{5280 \text{ ft/mile}}{3600 \text{ sec/hr}} = \text{ft,sec}$

Therefore, using the same 30,000 pound engine at 50 MPH, the momentum realized in pounds/second is:

M = m v = f v

$M = \dfrac{30,000}{32.2} v$

M = (932 slugs) v
M = (932 slugs)(73 ft/sec)
M = 68,036 lb/sec

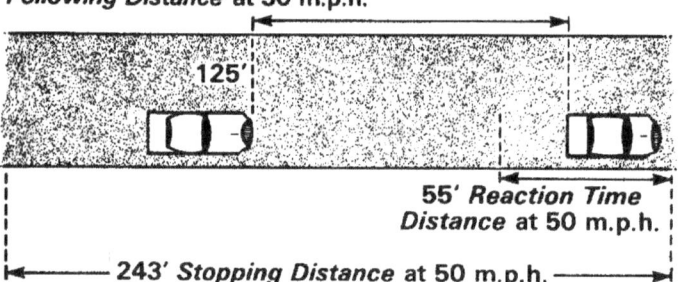

100' - 150' appropriate
Following Distance at 50 m.p.h.

125'

55' Reaction Time
Distance at 50 m.p.h.

243' *Stopping Distance* at 50 m.p.h.

O. Stopping Distance is the result of which two (2) other distances?

_____

2. _____

P. When calculating safe following distances, two (2) methods may be used to determine you are far enough behind the vehicle in front.

1. Describe the "2-Second Rule" _____

_____

_____

2. Describe the "3-Second Rule" _____

_____

_____

Q. Notes on Stopping Distance:

_____

_____

_____

_____

_____

R. Seat Belts (Passenger Restraints)

   A. Discuss the two (2) distinct collision types regarding unrestrained vehicle occupants:

      1. _____ 2. _____

   B. What crash dynamics occur during collisions?

   C. Discuss the importance of emergency responders wearing seat belts and your present company or departmental policy. How might it be improved or why is it sufficient.

Section VII-Vehicle Maintenance and Records

   A. List four (4) functions good maintenance records on Emergency Vehicles serve.

      1. _____

      2. _____

      3. _____

      4. _____

   B. Discuss your agency's present preventative maintenance program. How could it be bettered?

Checklists may be constructed for your own use from manufacturer's manuals after modification. Or as a greater asset, use VFIS Forms in your record system. (the following pages)

Section VIII-Vehicle Standard Operating Procedures

   A. Standard Operating Procedures (S.O.P.s) are a means by which administrative and operational policies are constructed. Discuss how your agency could utilize S.O.P.s for normal emergency operations.

   B. Nine (9) functions of S.O.P.s were discussed. Indicate five (5).

      1. _____

      2. _____

      _____

      4. _____

      5. _____

   C. What are the five (5) questions S.O.P.s answer in their construction? (This is the method journalists use in your daily newspaper.)

      1. _____

      2. _____

      3. _____

      4. _____

      5. _____

   D. How often should you review and change your agency's S.O.P.s? _____

_____

_____

Attach copies of your state's Vehicle Code pertaining to the duties of Emergency Vehicle Operators'

OUTDOOR TRAINING SESSION

We conclude the classroom portion with a detailed explanation of the hands-on driver training course which is excerpted from National Fire Protection Association Pamphlet 1002, Fire Apparatus Driver/Operator Professional Qualifications. From experience we have found that this can be set up to demonstrate many of the maneuvers that emergency service personnel will be expected to perform in vehicle operations. The following outline of the course and score sheet will be your personal record of how you have mastered the evolutions on this course. It can be used to judge your individual ability as well as a progress chart when you do the hands-on portion at future dates.

DRIVING OBSTACLE COURSE

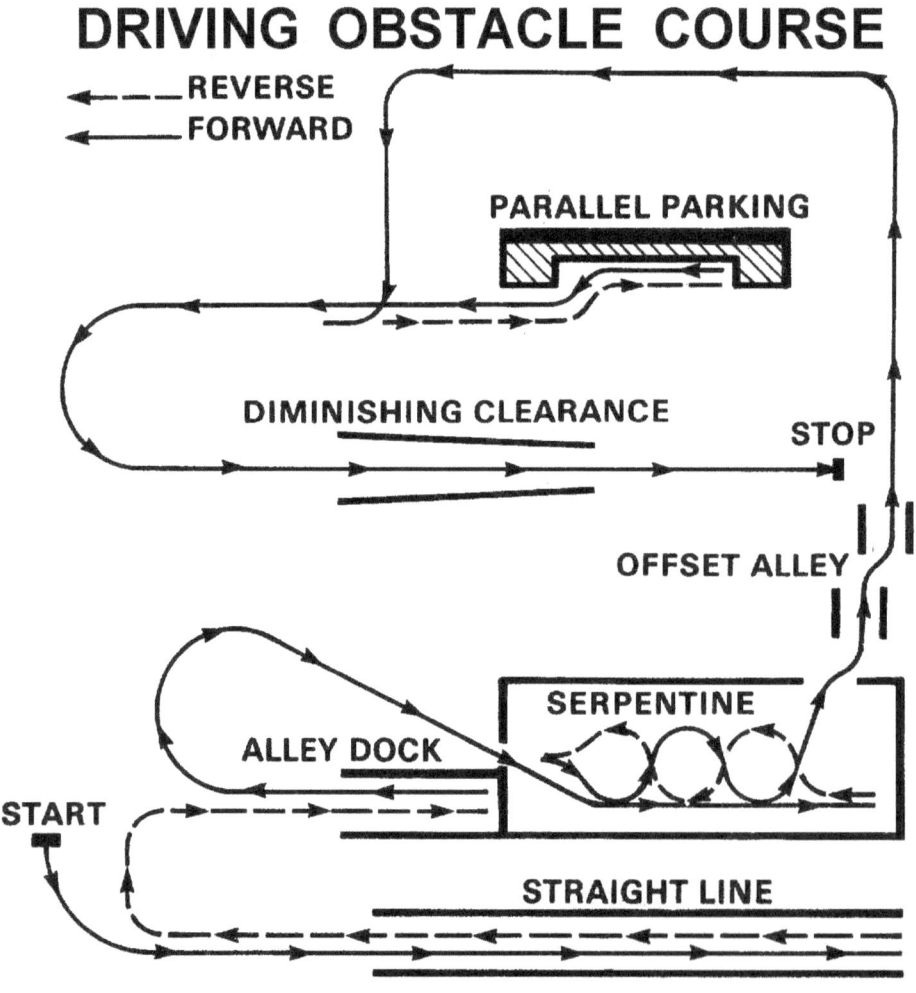

Volunteer Firemen's Insurance Services
WEEKLY EMERGENCY VEHICLE REPORT

PAGE NO. ———

NAME OF COMPANY ————————————————————————————

ADDRESS ————————————————————————————————

EMERGENCY VEHICLE MFG.: ——————————————————————————

YEAR ———————————— SERIAL NO.: ———————————— TYPE: ————————

REQUIRED TIRE PRESSURE: ——————————

DATE INSPECTION COMPLETED	INSPECTOR	BATTERY CHECK	BRAKING SYSTEM	ELECTRICAL SYSTEM, LIGHTS & SIRENS	TIRES & WHEEL LUGS	FUEL LEVEL	OIL LEVEL, ENG. & HYD.	HYDRAULIC SYSTEM	PUMP CHECK	COOLING SYSTEM	LUBRICATION PUMP & LADDER	ENGINE CHECK	BOOSTER, TANK LEVEL	DOORS— COMPARTMENT	PORTABLE EQUIPMENT	SPECIAL REMARKS ON ROAD TEST INSPECTION USE OTHER SIDE

REMARKS: *(Please itemize procedure taken on unsatisfactory inspection items noted on opposite side)*

INSPECTION DATE:	REPAIR DATE:	COMMENTS:	REPAIRS COMPLETED- BY:	DATE:

Volunteer Firemen's Insurance Services, Inc.
a subsidiary of the GLATFELTER INSURANCE GROUP

Printed in U.S.A.

Item no. C10.007 (Rev 7/85)

Volunteer Firemen's Insurance Services, Inc.

EMERGENCY VEHICLE ACCIDENT/LOSS INVESTIGATION REPORT

FIRE DEPARTMENT _____ DATE _____

ADDRESS _____

NAME OF DRIVER _____ VEHICLE IDENTIFICATION NUMBER _____

TYPE OF EMERGENCY SERVICE VEHICLE _____

DATE DRIVER LAST CERTIFIED ON ABOVE VEHICLE _____

DATE OF ACCIDENT _____ TIME _____ DATE REPORTED _____

LOCATION OF ACCIDENT _____

ROADWAY:
☐ Straight
☐ Curve
☐ On grade
☐ Level
☐ Hillcrest
☐ Dry
☐ wet
☐ Muddy
☐ Snowy
☐ Icy
☐ Oily

☐ P 2-lane
☐ 3-Lane
☐ 4-lane
☐ Divided
☐ Rural
☐ (Other) _____
☐ Lanes marked
☐ Lanes unmarked
☐ No road defects
☐ Holes, ruts, etc.
☐ Loose material
☐ (Other) _____

ACCIDENT OCCURRED:
☐ At station
☐ Responding to emergency
☐ At emergency scene
☐ Returning from emergency
☐ Training
☐ Convention or parade
☐ (Other) _____

TYPE OF LOSS
☐ Personal Injury
☐ Property damage

WEATHER
☐ Clear
☐ Rain
☐ Snow
☐ sleet
☐ Fog
☐ (Other) _____

DESCRIPTION OF ACCIDENT _____

_____

_____

_____

_____

_____

MOTOR VEHICLE DIAGRAM
Complete the following diagram showing direction & positions of automobiles involved, designating clearly point of contact.

Indicate North

GIVE STREET NAMES AND DIRECTIONS

YOUR VEH. ◀━━━ OTHER VEH. ◁━▷

INSTRUCTIONS:
1. Show vehicles and direction of travel.
2. Use solid line to show path of each vehicle before accident. ▭▷

dotted line after accident . . .

SAFETY OFFICER ANALYSIS:

WHAT ACTS. FAILURES TO ACT AND/OR CONDITIONS CONTRIBUTED MOST DIRECTLY TO THIS ACCIDENT? (IMMEDIATE CAUSE)

_____

_____

_____

WHAT ARE THE BASIC OR FUNDAMENTAL REASONS FOR THE EXISTENCE OF THESE ACTS AND/OR CONDITIONS? (FUNDAMENTAL CAUSE)

_____

_____

_____

WHAT ACTION HAS OR WILL BE TAKEN TO PREVENT RECURRENCE? PLACE "X" BY ITEMS COMPLETED

_____

_____

_____

_____

_____

_____

_____

_____

_____

_____

SAFETY OFFICER COMMENTS _____

_____

_____

_____

SIGNATURE OF SAFETY OFFICER _____ DATE _____

Volunteer Firemens Insurance Services, inc.

PERSONNEL FILE

(Attach Photo Here)

Address: _____

City or Town: _____ State: _____ Zip: _____

Telephone #: _____ (H) _____ (B)

Employer: _____

Address: _____

City/Town: _____ State: _____ Zip: _____

Social Security No.: _____ Driver License No.: _____

Married: _____ Year: _____ Spouse's Name: _____

Beneficiary: 1st: _____ 2nd: _____

Dependents: _____   _____

| Name | DOB | Name | DOB |

_____   _____

| Name | DOB | Name | DOB |

Date Joined Dept.: _____ Date Terminated: _____ Reason: _____

EQUIPMENT ISSUED

Item	Ser. # or Size	Date Iss.	Date Ret.

OFFICES HELD

Title	From	To	Remarks	By

INDIVIDUAL TRAINING RECORD

NAME

RANK

DATE	SUBJECT	LOCATION	INSTRUCTOR	HOURS THEORY	HOURS SKILL	SUB-TOTAL	YEARLY TOTAL

P.O. BOX 2726
York, PA 17405

Item No. C10-013(11/85)
Printed in U.S.A.

Volunteer Firemen's Insurance Services, Inc.

ANNUAL MEDICAL STATEMENT OF PERSONNEL

NOTE: Thus form is designed to provtde the officer in charge of all personnel a complete history of physical status as of the date indicated without the need for expense physical examinations. It is recommended that the form be completed on an annual basis by all drivers of emergency vehicles as Well as other active members. If any of the questions are answered "YES", be sure the answer is fully explained.

QUESTIONS:

NAME: _____

ADDRESS: _____

CITY & STATE _____ ZIP: _____

FULL TIME OCCUPATION: _____

NAME OF ORGANIZATION: _____

ARE YOU A: ☐ Certified Vehicle Driver'   ☐ Driver Trainee

Social Security No. _____
What is your Valid Slate Operators Plate No. _____

---

1. Birth Date: Month. _____ Day: _____ Year: _____

2. Eyesight:
a. Have you lost use ot either eye'? _____ R _____ L ....   a. ☐ Yes ☐ No
b. Is peripheral (side) vtsion restricted? ....   b. ☐ Yes ☐ No
c. Are you color blind? ....   c. ☐ Yes ☐ No
d. Do you have, or have you ever had, cataracts? ....   d. ☐ Yes ☐ No
e. Are actual deficiencies corrected by glasses or contact lenses?   e. ☐ Yes ☐ No
f. Date of last eye examination. ....   f. _____

3. Hearing:
a. Do you have difficulty hearing normal conversation level? ....   a. ☐ Yes ☐ No
b. Do you use a hearing aid? ....   b. ☐ Yes ☐ No

4. Diabetes:
a. Have you ever been treated for diabetes? ....   a. ☐ Yes ☐ No
b. Describe current medication and dosage, if any, and method of administration under "remarks"
c. Date of latest blood sugar test: ....   c. _____

5. Heart:
a. Have you ever been treated for heart disease? ....   a. ☐ Yes ☐ No
b. Describecondition: ....   _____
c. Describe current medication and dosage, if any, under
   "remarks"
d. Do you have a pacemaker? ....   d. ☐ Yes ☐ No
e. Date of last treatment or check-up: ....   e. _____

6. Epilepsy:
a. Have you ever been treated for epilepsy? ....   a. ☐ Yes ☐ No
b. If "Yes". when was your last seizure? ....   b. _____
c. Describe current medication and dosage, if any, under
   "remarks"

REMARKS:

NOTE: II any question is answered "YES". give particulars below. For medical histories, underline the item and identify by referring to question number and letter. Give deter, symptoms, duration. treatment results, names and addresses of doctors, hospitals, etc.

QUESTIONS:

7. **Blood Pressure:**
   a. Have you ever been treated for high blood pressure? .... a. ☐ Yes ☐ No
   b. If "Yes", when were you treated? .... b. _____
   c. What was your last reading? .... c. _____
   d. Describe current medication and dosage, if any, under "remarks".

8. **Limbs:**
   a. Have you lost an arm or leg? .... a. ☐ Yes ☐ No
   b. Have you lost the use of an arm or a leg? .... b. ☐ Yes ☐ No
   c. Does vehicle have special controls? .... c. ☐ Yes ☐ No
   d. If "Yes" to any of the above, describe under "remarks".

9. **Miscellaneous:**
   a. Have you ever had, or been treated for, Convulsions? .... a. ☐ Yes ☐ No
   b. If "Yes", give date of last treatment and describe current medication and dosage, if any, under "remarks".
   c. Have you ever had any Fainting Spells? .... c. ☐ Yes ☐ No
   d. If "Yes", give date last treatment and describe current medication and dosage, if any, under "remarks".
   e. Have you ever had, or been treated for, Loss of Equilibrium? .... e. ☐ Yes ☐ No
   f. If "Yes", give date of last treatment and describe current medication and dosage, if any, under "remarks".
   g. Have you ever been treated for Alcohol or Drug Abuse? .... g. ☐ Yes ☐ No
   h. If "Yes", give date of last treatment and describe current medication and dosage, if any, under "remarks".
   i. Have you ever been treated for Mental Illness? .... i. ☐ Yes ☐ No
   j. If "Yes", give date of last treatment and describe current medication and dosage, if any, under "remarks".

10. **What is the date of your last physical examination?** .... _____

11. An there any restrictions posted on your vehicle operator's license? ... ☐ Yes ☐ No

12. Are you under the can of a physician for any condition not mentioned above which may affect your ability to operate a motor vehicle? .... ☐ Yes ☐ No

13. **When and for what purpose, did you last consult a doctor?**

   _____

   _____

14. **FULL NAME, address and telephone number of your personal physician.**

   NAME: _____

   ADDRESS: _____

   CITY & STATE: _____ ZIP: _____

   PHONE NO.: _____

REMARKS:

The answer to the to the above are complete, accurate and true to the best of my knowledge.

_____   _____
SIGNATURE OF PERSON NAMED ABOVE   DATE

AUTHORIZATION FOR RELEASE OF INFORMATION

_____   _____
SIGNATURE OF PERSON NAMED ABOVE   DATE

Volunteer Firemen's Insurance Services
a subsidiary of the GLATFELTER INSURANCE GROUP

Item No. C10:009
Rev. 10/85

DATE: _____

NAME OF ORGANIZATION: _____

OPERATOR'S NAME: _____

ADDRESS: _____

CITY & STATE: _____ ZIP: _____

ANNUAL MEDICAL STATEMENT COMPLETED _____
                                                    (DATE)

VALID OPERATOR'S PLATE NO: _____

R O A D   T E S T   M I L E S: SOCIAL SECURITY NO: _____
        (NOTE Recommended   minimum 15 miles)

TESTED ON:

☐ IPUMPER   ☐ TANKER
☐ AERIAL APPARATUS   ☐ RESCUE TRUCK
☐ AMBULANCE   OTHER: _____

EMERGENCY OPERATOR CERTIFICATION TEST

☐ DESIGNATED OPERATOR
☐ TRAINEE

CHECK APPROPRIATE BOXES

	Yes	No	Satisfactory	Unsatisfactory	REMARKS
1. Operator has been thoroughly familiarized with state and local laws governing Fire Department and Emergency Service vehicle operation:	☐	☐	☐	☐	

Pre-trip inspection:

	Yes	No	Satisfactory	Unsatisfactory	REMARKS
2. Operator has received personal instructions in the care end use of the following equipment:					
Service brakes, including trailer brake connections. (Brake airline hoses, compressor belts. tractor protection valve, air pressure)	☐	☐	☐	☐	
Parking(hand)brake	☐	☐	☐	☐	
Steering mechanism :	☐	☐	☐	☐	
Lighting devices and reflector. (Headlights-high and low beam, clearance and identification lights, tail lights, turn signals, reflectors, side markers, four-way flasher, cab lights)	☐	☐	☐	☐	
Tires. (Inflation. tread wear, cuts in sidewalls, lugs, or studs, grease leaks around hubs, mud flaps, valve caps, spare tire)	☐	☐	☐	☐	
Horns, sirens and/or audible signals	☐	☐	☐	☐	
Windshield wiper(s)	☐	☐	☐	☐	
Rear-vision mirrors. (Clean,properlyadjusted)	☐	☐	☐	☐	
Coupling devices. (Fifth wheel, jaws, release lever on pintle hook, tow-bar, safety chains, converter gear,airlines	☐	☐	☐	☐	

Operating performance:

	Yes	No	Satisfactory	Unsatisfactory	REMARKS
3. Operator has been instructed In operation of vehicle according to standard safety procedures:					
(Checks air pressure and instruments, emergency brake set, disengages clutch, warms up engine, proper gear selection, checks traffic, shifts gears smoothly)	☐	☐	☐	☐	
4. Operator has been trained In use of vehicle controls and emergency equipment:					
(Clutch and transmission. brakes, steering, lights, jacks. tools, tire chains, emergency warning devices. fire extinguisher)	☐	☐	☐	☐	

CERTIFICATION TEST Continued

5. Operator has had at least fifteen (15) miles of driving experience in traffic, demonstrating ability in following:

(Leaving curb, speed control, smoothness of operation, shifting gears, anticipates traffic problems. obeys traffic laws, signals properly, allows sufficient passing room, passes cautiously and smoothly, uses mirror): ....

☐ Yes	☐ No	☐ Satisfactory	☐ Unsatisfactory	REMARKS
☐	☐	☐	☐	

6. Operator has received training in turning vehicles and demonstrated ability to do so:

(Signals well in advance, turns from proper lane, looks all around before turning, ' turns at proper speed, turns into proper lane. yields right-of-way) ....

| ☐ | ☐ | ☐ | ☐ | |

7. Operator has been given instructions in braking and slowing vehicle by means other than braking:

(Signals, checks mirrors, smoothness, proper use of tractor protection valve, uses engine to reduce speed by shifting to lower gear ....

| ☐ | ☐ | ☐ | ☐ | |

8. Operator has displayed skills in backing and parking vehicles:

(Gets out and checks, sounds horn when necessary, avoids backing to blind side, backs slowly, parks off pavement, secures unit properly, uses guide if necessary).

| ☐ | ☐ | ☐ | ☐ | |

9. Operator has demonstrated proficiency in driving emergency vehicle obstacle course:

I certify that the above named operator was given a road test using an approved road test form, under my supervision on the date specified below. It is my considered opinion that this driver possesses sufficient operating skill to safely operate the type of emergency motor vehicles checked above.

Signature of examiner: _____

Title: _____

Organization: _____

Address _____

City & State: _____ ZIP: _____

Date: _____

VOLUNTEER FIREMAN'S INSURANCE SERVICES, INC.
P.O. BOX 2726 · YORK, PENNSYLVANIA 17405

FORM NO. 5174-363
PRINTED IN U S A.
COPYRIGHT 1977 by V F IS.
C10:008

CP-3864 Ptd. In U.S.A.

Volunteer Firemen's
Insurance Services, Inc.
. . . a subsidiary of the GLATFELTER INSURANCE GROUP

DRIVER'S OBSTACLE COURSE

This obstacle course is designed to measure the skills of drivers of emergency vehicles. Through its use, training officials can determine the progress each trainee has made over a given period of time. The "Recommended Time" allocated to each vehicle type indicates an ideal score toward which trainees may use as their objective. It may also be used to test present drivers' skills against a norm. The obstacle course is planned to duplicate seven situations in which driver skill, judgement and knowledge of the limitations of his vehicle are required for effective maneuvering. This course of driving tests is listed in the N.F.P.A. Publication #1002 titled FIRE APPARATUS DRIVER/OPERATOR PROFESSIONAL QUALIFICATIONS, 1976, in Appendix A. Scoring is based on total time required to complete the course plus the penalties assigned for mis-maneuvers.

NOTE: CREW MAY ASSIST DRIVER IN ALL OBSTACLES EXCEPT STOP SIGN NO. 7.

NAME: _____

COMPANY: _____

VEHICLE: _____

PENALTY SCHEDULE

OBSTACLE NO.	DESCRIPTION	ERROR	PENALTY
No. 1	Straight Line	Each cone brushed, moved or overturned Crossing any line, each time	10 sec. 3 sec.
No. 2	Alley Dock	Each cone brushed, moved or overturned Crossing any line, each time Stopping 18" or more short of dock stop Stopping 12"-17" short of dock stop Stopping 6"-11" short of dock stop	10 sec. 3 sec. 10 sec. 6 sec. 3 sec.
No. 3	Serpentine	Each pilon brushed, moved or overturned Failure to stop in time, either end of course Crossing any line, each time	10 sec. 10 sec. 3 sec.
No. 4	Offset Alley	Each cone brushed, moved or overturned Crossing any line. each time	10 sec. 3 sec.
No. 5	Parallel Parking	Each cone brushed, moved or overturned Crossing any line, each time If distance from curb line is 12" or more	10 sec. 3 sec. 3 sec.
No. 6	Diminishing Clearance	Each cone brushed, moved or overturned Crossing any line, each time	10 sec. 3 sec.
No. 7	Stop Sign	Crossing stop line Stopping 18" or more short of line Stopping 12" to 17" short of line Stopping 6" to 11" short of line	10 sec. 10 sec. 6 sec. 3 sec.

SCORE CARD

OBSTACLE NO.:	RUN NO. 1	RUN NO. 2	RUN NO. 3	RUN NO. 4	RUN NO. 5	RUN NO. 6
	Date:	Date:	Date:	Date:	Date:	Date:
1.						
2.						
3.						
4.						
5.						
6.						
7.						
TOTAL PENALTIES +						
DRIVING TIME						
SCORE:						
INITIALS OF SCOREKEEPER:						

DRIVING OBSTACLE COURSE

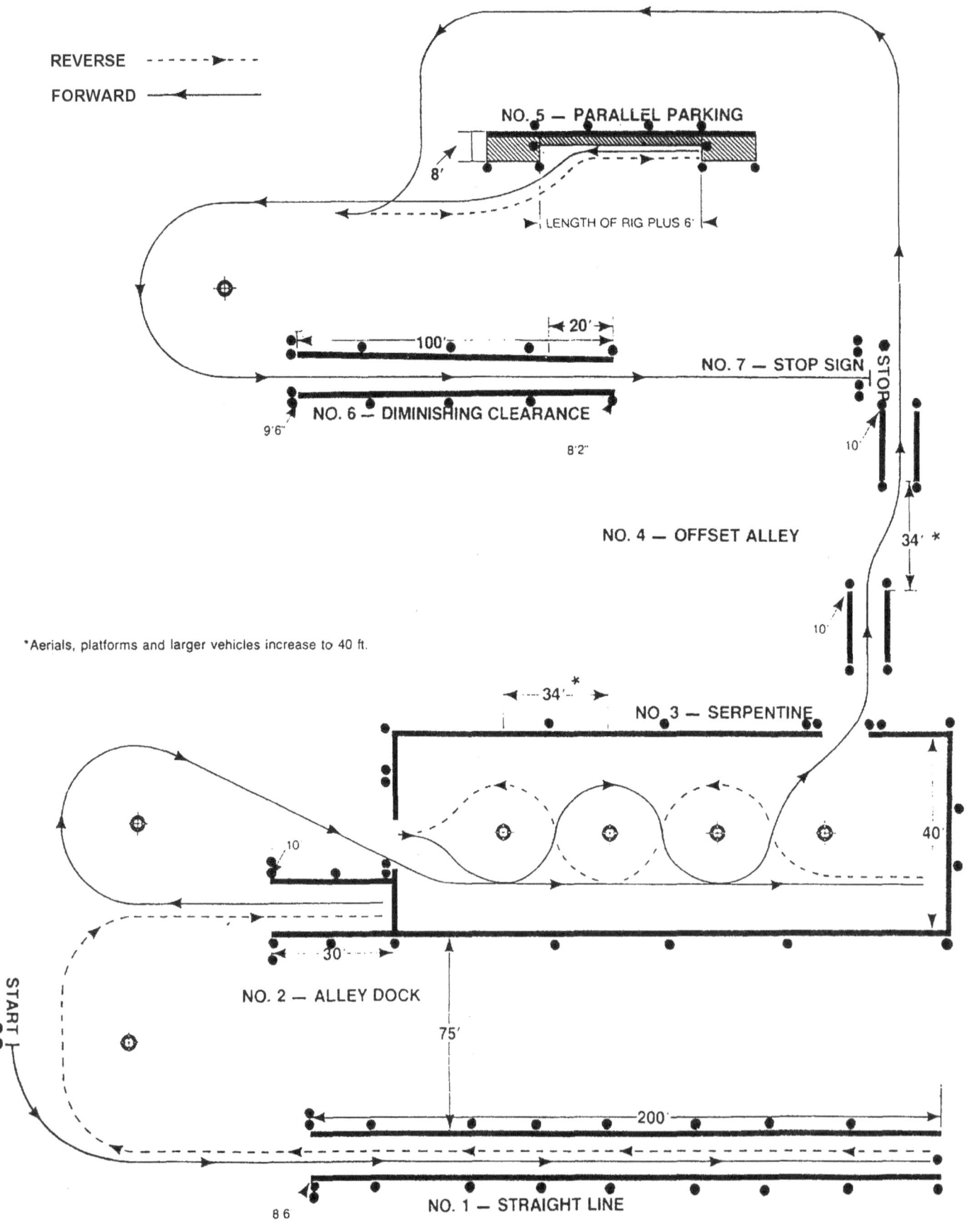

REVERSE ------►

FORWARD ◄——

NO. 5 — PARALLEL PARKING

8'

LENGTH OF RIG PLUS 6'

100'   20'

NO. 7 — STOP SIGN   STOP

NO. 6 — DIMINISHING CLEARANCE

9'6"   8'2"   10'

NO. 4 — OFFSET ALLEY

34' *

*Aerials, platforms and larger vehicles increase to 40 ft.   10'

34' *   NO. 3 — SERPENTINE

40'

10'

NO. 2 — ALLEY DOCK

30'   75'

START   200'

NO. 1 — STRAIGHT LINE

8 6